This report contains the collective views of an international group of experts and does not necessarily represent the decisions or the stated policy of the United Nations Environment Programme, the International Labour Organisation, or the World Health Organization.

Environmental Health Criteria 182

THALLIUM

First draft prepared by Professor G. Schaub,
Institute for Zoology and Parasitology,
Ruhr University, Bochum, Germany

Published under the joint sponsorship of
the United Nations Environment Programme,
the International Labour Organisation,
and the World Health Organization

World Health Organization
Geneva, 1996

The **International Programme on Chemical Safety (IPCS)** is a joint venture of the United Nations Environment Programme, the International Labour Organisation, and the World Health Organization. The main objective of the IPCS is to carry out and disseminate evaluations of the effects of chemicals on human health and the quality of the environment. Supporting activities include the development of epidemiological, experimental laboratory, and risk-assessment methods that could produce internationally comparable results, and the development of manpower in the field of toxicology. Other activities carried out by the IPCS include the development of know-how for coping with chemical accidents, coordination of laboratory testing and epidemiological studies, and promotion of research on the mechanisms of the biological action of chemicals.

WHO Library Cataloguing in Publication Data

Thallium.

(Environmental health criteria ; 182)

1.Thallium - toxicity I.Series

ISBN 92 4 157182 9 (NLM Classification: QV 618)
ISSN 0250-863X

The World Health Organization welcomes requests for permission to reproduce or translate its publications, in part or in full. Applications and enquiries should be addressed to the Office of Publications, World Health Organization, Geneva, Switzerland, which will be glad to provide the latest information on any changes made to the text, plans for new editions, and reprints and translations already available.

The designations employed and the presentation of the material in this publication do not imply the expression of any opinion whatsoever on the part of the Secretariat of the World Health Organization concerning the legal status of any country, territory, city or area or of its authorities, or concerning the delimitation of its frontiers or boundaries.

The mention of specific companies or of certain manufacturers' products does not imply that they are endorsed or recommended by the World Health Organization in preference to others of a similar nature that are not mentioned. Errors and omissions excepted, the names of proprietary products are distinguished by initial capital letters.

PRINTED IN FINLAND
96/10845 – VAMMALA – 5500

CONTENTS

ENVIRONMENTAL HEALTH CRITERIA FOR THALLIUM

NOTE TO READERS OF THE CRITERIA MONOGRAPHS

Every effort has been made to present information in the criteria monographs as accurately as possible without unduly delaying their publication. In the interest of all users of the Environmental Health Criteria monographs, readers are requested to communicate any errors that may have occurred to the Director of the International Programme on Chemical Safety, World Health Organization, Geneva, Switzerland, in order that they may be included in corrigenda.

* * *

A detailed data profile and a legal file can be obtained from the International Register of Potentially Toxic Chemicals, Case postale 356, 1219 Châtelaine, Geneva, Switzerland (Telephone No. 9799111).

* * *

This publication was made possible by grant number 5 U01 ES02617-15 from the National Institute of Environmental Health Sciences, National Institutes of Health, USA, and by financial support from the European Commission.

Environmental Health Criteria

P R E A M B L E

Objectives

In 1973 the WHO Environmental Health Criteria Programme was initiated with the following objectives:

(i) to assess information on the relationship between exposure to environmental pollutants and human health, and to provide guidelines for setting exposure limits;

(ii) to identify new or potential pollutants;

(iii) to identify gaps in knowledge concerning the health effects of pollutants;

(iv) to promote the harmonization of toxicological and epidemiological methods in order to have internationally comparable results.

The first Environmental Health Criteria (EHC) monograph, on mercury, was published in 1976 and since that time an ever-increasing number of assessments of chemicals and of physical effects have been produced. In addition, many EHC monographs have been devoted to evaluating toxicological methodology, e.g., for genetic, neurotoxic, teratogenic and nephrotoxic effects. Other publications have been concerned with epidemiological guidelines, evaluation of short-term tests for carcinogens, biomarkers, effects on the elderly and so forth.

Since its inauguration the EHC Programme has widened its scope, and the importance of environmental effects, in addition to health effects, has been increasingly emphasized in the total evaluation of chemicals.

The original impetus for the Programme came from World Health Assembly resolutions and the recommendations of the 1972 UN Conference on the Human Environment. Subsequently the work became an integral part of the International Programme on Chemical Safety (IPCS), a cooperative programme of UNEP, ILO and WHO. In this manner, with the strong support of the new partners, the importance of occupational health and environmental effects was fully recognized. The EHC monographs have become widely established, used and recognized throughout the world.

The recommendations of the 1992 UN Conference on Environment and Development and the subsequent establishment of the Intergovernmental Forum on Chemical Safety with the priorities for action in the six programme areas of Chapter 19, Agenda 21, all lend further weight to the need for EHC assessments of the risks of chemicals.

Scope

The criteria monographs are intended to provide critical reviews on the effect on human health and the environment of chemicals and of combinations of chemicals and physical and biological agents. As such, they include and review studies that are of direct relevance for the evaluation. However, they do not describe *every* study carried out. Worldwide data are used and are quoted from original studies, not from abstracts or reviews. Both published and unpublished reports are considered and it is incumbent on the authors to assess all the articles cited in the references. Preference is always given to published data. Unpublished data are only used when relevant published data are absent or when they are pivotal to the risk assessment. A detailed policy statement is available that describes the procedures used for unpublished proprietary data so that this information can be used in the evaluation without compromising its confidential nature (WHO (1990) Revised Guidelines for the Preparation of Environmental Health Criteria Monographs. PCS/90.69, Geneva, World Health Organization).

In the evaluation of human health risks, sound human data, whenever available, are preferred to animal data. Animal and *in vitro* studies provide support and are used mainly to supply evidence missing from human studies. It is mandatory that research on human subjects is conducted in full accord with ethical principles, including the provisions of the Helsinki Declaration.

The EHC monographs are intended to assist national and international authorities in making risk assessments and subsequent risk management decisions. They represent a thorough evaluation of risks and are not, in any sense, recommendations for regulation or standard setting. These latter are the exclusive purview of national and regional governments.

Content

The layout of EHC monographs for chemicals is outlined below.

- Summary - a review of the salient facts and the risk evaluation of the chemical
- Identity - physical and chemical properties, analytical methods
- Sources of exposure
- Environmental transport, distribution and transformation
- Environmental levels and human exposure
- Kinetics and metabolism in laboratory animals and humans
- Effects on laboratory mammals and *in vitro* test systems
- Effects on humans
- Effects on other organisms in the laboratory and field
- Evaluation of human health risks and effects on the environment
- Conclusions and recommendations for protection of human health and the environment
- Further research
- Previous evaluations by international bodies, e.g., IARC, JECFA, JMPR

Selection of chemicals

Since the inception of the EHC Programme, the IPCS has organized meetings of scientists to establish lists of priority chemicals for subsequent evaluation. Such meetings have been held in: Ispra, Italy, 1980; Oxford, United Kingdom, 1984; Berlin, Germany, 1987; and North Carolina, USA, 1995. The selection of chemicals has been based on the following criteria: the existence of scientific evidence that the substance presents a hazard to human health and/or the environment; the possible use, persistence, accumulation or degradation of the substance shows that there may be significant human or environmental exposure; the size and nature of populations at risk (both human and other species) and risks for environment; international concern, i.e. the substance is of major interest to several countries; adequate data on the hazards are available.

If an EHC monograph is proposed for a chemical not on the priority list, the IPCS Secretariat consults with the Cooperating Organizations and all the Participating Institutions before embarking on the preparation of the monograph.

Procedures

The order of procedures that result in the publication of an EHC monograph is shown in the flow chart. A designated staff member of IPCS, responsible for the scientific quality of the document, serves as Responsible Officer (RO). The IPCS Editor is responsible for layout and language. The first draft, prepared by consultants or, more usually, staff from an IPCS Participating Institution, is based initially on data provided from the International Register of Potentially Toxic Chemicals, and reference data bases such as Medline and Toxline.

The draft document, when received by the RO, may require an initial review by a small panel of experts to determine its scientific quality and objectivity. Once the RO finds the document acceptable as a first draft, it is distributed, in its unedited form, to well over 150 EHC contact points throughout the world who are asked to comment on its completeness and accuracy and, where necessary, provide additional material. The contact points, usually designated by governments, may be Participating Institutions, IPCS Focal Points, or individual scientists known for their particular expertise. Generally some four months are allowed before the comments are considered by the RO and author(s). A second draft incorporating comments received and approved by the Director, IPCS, is then distributed to Task Group members, who carry out the peer review, at least six weeks before their meeting.

The Task Group members serve as individual scientists, not as representatives of any organization, government or industry. Their function is to evaluate the accuracy, significance and relevance of the information in the document and to assess the health and environmental risks from exposure to the chemical. A summary and recommendations for further research and improved safety aspects are also required. The composition of the Task Group is dictated by the range of expertise required for the subject of the meeting and by the need for a balanced geographical distribution.

The three cooperating organizations of the IPCS recognize the important role played by nongovernmental organizations. Representatives from relevant national and international associations may be invited to join the Task Group as observers. While observers may provide a valuable contribution to the process, they can only speak at the invitation of the Chairperson.

EHC PREPARATION FLOW CHART

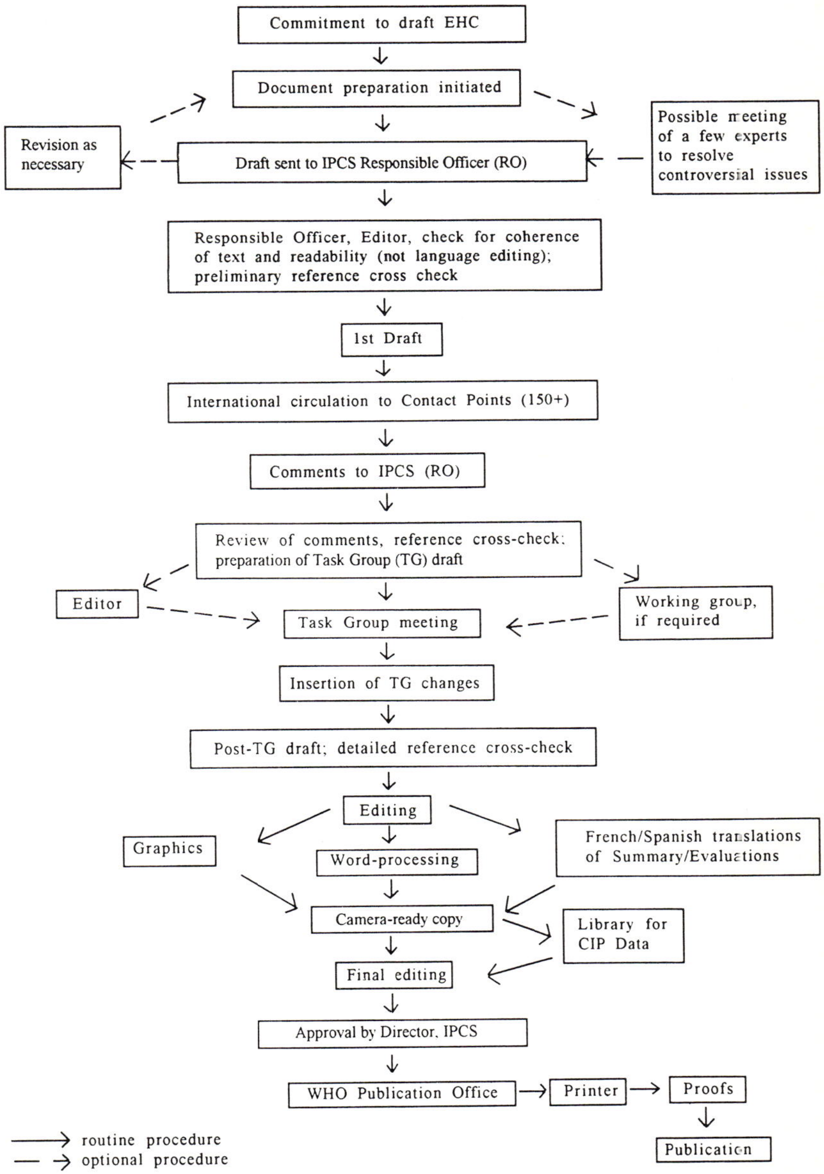

Observers do not participate in the final evaluation of the chemical; this is the sole responsibility of the Task Group members. When the Task Group considers it to be appropriate, it may meet *in camera*.

All individuals who as authors, consultants or advisers participate in the preparation of the EHC monograph must, in addition to serving in their personal capacity as scientists, inform the RO if at any time a conflict of interest, whether actual or potential, could be perceived in their work. They are required to sign a conflict of interest statement. Such a procedure ensures the transparency and probity of the process.

When the Task Group has completed its review and the RO is satisfied as to the scientific correctness and completeness of the document, it then goes for language editing, reference checking, and preparation of camera-ready copy. After approval by the Director, IPCS, the monograph is submitted to the WHO Office of Publications for printing. At this time a copy of the final draft is sent to the Chairperson and Rapporteur of the Task Group to check for any errors.

It is accepted that the following criteria should initiate the updating of an EHC monograph: new data are available that would substantially change the evaluation; there is public concern for health or environmental effects of the agent because of greater exposure; an appreciable time period has elapsed since the last evaluation.

All Participating Institutions are informed, through the EHC progress report, of the authors and institutions proposed for the drafting of the documents. A comprehensive file of all comments received on drafts of each EHC monograph is maintained and is available on request. The Chairpersons of Task Groups are briefed before each meeting on their role and responsibility in ensuring that these rules are followed.

WHO TASK GROUP ON ENVIRONMENTAL HEALTH CRITERIA FOR THALLIUM

Members

Professor M. Balali-Mood, Poison Control Centre, Imam Reza Hospital, Mashhad University of Medical Sciences, Mashhad, Islamic Republic of Iran

Dr P. Doyle, Chemicals Evaluation Division, Environment Canada, Ottawa, Ontario, Canada

Professor G. Kazantzis, Imperial College of Science, Technology and Medicine, Centre for Environmental Technology, Royal School of Mines, London, United Kingdom (*Joint Rapporteur*)

Dr M. Kiilunen, Department of Industrial Hygiene & Toxicology, Institute of Occupational Health, Helsinki, Finland

Mr H. Malcolm, Institute of Terrestrial Ecology, Monks Wood Experimental Station, Huntingdon, Cambridgeshire, United Kingdom

Dr G. Nordberg, Department of Environmental Hygiene, Umea University, Umea, Sweden (*Chairman*)

Professor G. Schaub, Department of Zoology, Institute for Zoology and Parasitology, Ruhr University, Bochum, Germany (*Joint Rapporteur*)

Dr S. Velazquez, Environmental Criteria and Assessment Office, US Environmental Protection Agency, Cincinnati, Ohio, USA

Representatives of other organizations

Dr P. Montuschi, Department of Pharmacology, Catholic University of the Sacred Heart, Rome, Italy (representing the International Union of Toxicology)

Observers

Dr R. Cornelis, Institute for Nuclear Sciences, State University of Gent, Gent, Belgium

Secretariat

Dr P.G. Jenkins, International Programme on Chemical Safety, World Health Organization, Geneva, Switzerland (*Secretary*)

ENVIRONMENTAL HEALTH CRITERIA FOR THALLIUM

A WHO Task Group on Environmental Health Criteria for Thallium met in Geneva from 12 to 16 December 1994. Dr P.G. Jenkins, IPCS, welcomed the participants on behalf of Dr M. Mercier, Director of the IPCS, and the three IPCS cooperating organizations (UNEP/ILO/WHO). The Group reviewed and revised the draft and made an evaluation of the risks for human health and the environment from exposure to thallium.

The first draft was prepared by Professor G. Schaub, Institute for Zoology and Parasitology, Ruhr University, Bochum, Germany. He also prepared the second draft, incorporating comments received following circulation of the first draft to the IPCS contact points for Environmental Health Criteria monographs.

Dr P.G. Jenkins, IPCS, was responsible for both the overall scientific content and the technical editing.

The efforts of all who helped in the preparation and finalization of the monograph are gratefully acknowledged.

ABBREVIATIONS

AAS	atomic absorption spectrometry
AES	atomic emission spectrometry
AMP	amperometric titration
CRMs	certified reference materials
DPASV	differential pulse anodic stripping voltametry
EDL	electrode discharge lamp
EDTA	ethylenediaminetetraacetic acid
GABA	gamma-aminobutyric acid
GDMS	glow discharge mass spectrometry
GFAAS	graphite furnace atomic absorption spectrometry
GLP	good laboratory practice
ICP	inductively coupled plasma
IDMS	isotope dilution mass spectrometry
LOEL	lowest-observed-effect level
MED	minimum effective dose
MIBK	methyl isobutyl ketone
MS	mass spectrometry
NAA	neutron activation analysis
NaDDC	sodium diethyldithiocarbamate
NADP	nicotinamide adenine dinucleotide phosphate
NOEL	no-observed-effect level
PAA	photon activation analysis
TLV	threshold limit value
tRNA	transfer ribonucleic acid

1. SUMMARY

1.1 Identity, physical and chemical properties, and analytical methods

Elemental thallium is a soft and malleable metal with a bluish-white colour. When exposed to humid air or water, thallium is oxidized rapidly on the surface or the hydroxide is formed, respectively. Thallium has two important oxidation states, thallium(I) and thallium(III). Monovalent (thallous) compounds behave like alkali metals, e.g. potassium, whereas the trivalent (thallic) compounds are less basic, resembling aluminium. In contrast to inorganic compounds in which the thallium(I) ion is more stable in aqueous solutions than the thallium(III) ion, the latter is more stable in organic compounds.

The determination of thallium in environmental samples is somewhat difficult as concentrations are in the μg/kg range or less. Generally the limits of determination for minerals, soils and dusts are about 20 μg/kg, for aqueous solutions about 0.1 μg/litre, and for biological materials a few μg/kg, when no pre-concentration of thallium is applied.

Graphite furnace atomic absorption spectrometry (GFAAS) is an analytical method well suited for applications where high sensitivity is required from small sample amounts with thallium present at concentrations of a few μg/kg. Isotope dilution mass spectrometry (IDMS) and inductively coupled plasma-mass spectrometry (ICP-MS), possibly combined with isotope dilution, are excellent methods for determinations offering good precision and accuracy at the μg/kg level.

1.2 Sources of human and environmental exposure

Thallium is present in the environment as a result of natural processes and from man-made sources. It is ubiquitous in nature and occurs especially in sulfide ores of various heavy metals, but normally in low concentrations. There are only a few areas with a naturally very high thallium concentration.

Thallium is produced industrially only in small quantities (the worldwide industrial consumption in 1991 was 10-15 tonnes/year). Thallium and its compounds have a wide variety of industrial uses. Its uses as a depilatory agent for humans and as a rodenticide and insecticide are now severely restricted. The main uses are in the electrical and electronic industries and in the production of special glasses. Another important field of application is the use of radioisotopes in medicine for scintigraphy and the diagnosis of melanoma and the use of arylthallium(III) compounds in biochemistry.

Losses to the environment mainly occur from mineral smelters (deposits of waste material and emissions into the atmosphere), coal-burning power-generating plants, brickworks, and cement plants (all emissions into the atmosphere). From 2000 to 5000 tonnes/year are estimated to be mobilized world-wide by industrial processes. Emissions of thallium from industrial processes vary widely according to the type of industry.

Emissions from coal-fired power-generating plants can contain a thallium concentration of 700 $\mu g/m^3$ exhaust air and those from cement plants up to 2500 $\mu g/m^3$. The latter value may be reduced to < 25 $\mu g/m^3$ by using other raw materials and changing the production process. Thallium volatilizes during the burning of coal or raw material for cement production and recondenses on the surface of ash particles in cooler parts of the system. These particles contain up to 50 mg thallium/kg fly-ash and are often of small size, so that only 50% of them are held back by filters in cement plants. Also, about one third of emitted particles from power-generating plants are of the small particle size which can be deposited in the lower respiratory tract.

Effluent from mine tailing ponds containing up to 1620 and 36 μg/litre caused elevated levels of 88 and 1 μg/litre, respectively, in connecting rivers. Rainwater ponds around a cement plant contained up to 37 μg/litre. In soil maximal concentrations of 60 mg/kg have been found near waste materials from mines; 2, 0.6 and 27 mg/kg have been found in the vicinity of base metal smelters, brickworks and cement plants, respectively.

In contaminated areas the majority of vegetables, fruits and meat contain less than 1 mg thallium/kg fresh weight. Concentrations are higher in cabbages (Brassicaceae), with up to 45 mg/kg reported in green kale. Concentrations of thallium in the tissues of farm animals correlate with concentrations in the fodder. In the vicinity of some cement plants, increased concentrations in fodder (e.g., up to 1000 mg/kg in rape) and beef and rabbit meat (up to 1.5 and 5.8 mg/kg, respectively) have been reported.

1.3 Environmental transport, distribution and transformation

Near point sources such as coal-fired power-generating stations, some cement plants and metal smelting operations, the major source of thallium in air is emission of fly ash. The results of one study indicate that nearly all of the thallium in fly dust from a cement plant was present as soluble thallium(I) chloride.

The fate of thallium added to soil (in deposited fly ash, for example) depends largely on soil type. Retention will be greatest in soils that contain large amounts of clay, organic matter and iron/manganese oxides. Incorporation into stable complexes causes enhanced thallium concentrations only in the upper levels of soils. The uptake of thallium by vegetation increases as soil pH decreases. In some strongly acid soils significant amounts of thallium can be leached to local ground and surface water.

Most dissolved thallium in freshwater is expected to be in the monovalent form. However, in strongly oxidized fresh water and most seawater trivalent thallium may predominate. Thallium can be removed from the water column and accumulate in sediment by various exchange, complexation or precipitation reactions.

Although thallium can bioconcentrate, it is not likely to biomagnify in aquatic or terrestrial food webs.

1.4 Environmental levels and human exposure

In areas not contaminated by thallium, concentrations in air are usually < 1 ng/m^3, those in water < 1 μg/litre, and those in water sediments < 1 mg/kg. Mean concentrations in the earth's crust range from 0.1 to 1.7 mg/kg, but very high concentrations are possible, e.g., in coal up to 1000 mg/kg, and the rarely found minerals of thallium consist of up to 60% of the element. Food of plant and animal origin usually contains < 1 mg/kg dry weight and the human average dietary intake of thallium appears to be less than 5 μg/day. Uptake via the respiratory system is estimated to be < 0.005 μg thallium/day.

There are only limited data about the actual thallium content of workplace air. The most recent (1980s) concentrations of thallium observed were < 22 μg thallium/m^3 (in the production of a special thallium alloy and in a thallium smelter). Average urinary concentrations were determined to be in the range of 0.3-8 μg/litre for cement workers and 0.3-10.5 μg/litre for foundry workers.

1.5 Kinetics and metabolism in laboratory animals and humans

Thallium is rapidly and well absorbed through the gastrointestinal and respiratory tracts and is also taken up through the skin. It is rapidly distributed to all organs and passes the placenta (as indicated by the rapid fetal uptake) and the blood-brain barrier. Because of its rapid accumulation in cells, concentrations of thallium in whole blood do not reflect the levels in tissues. In acute poisoning of experimental animals or humans, initially high concentrations of thallium appear in the kidney, low concentrations in fat tissue and brain, and intermediate concentrations in the other organs; later the thallium concentration of the brain also increases.

Elimination of thallium may occur through the gastrointestinal tract (mainly by mechanisms independent of biliary excretion), kidney, hair, skin, sweat and breast milk. Intestinal reabsorption (mainly from the colon) may occur with a consequent decrease in total body clearance. In rats, the main routes of thallium elimination are gastrointestinal (about two thirds) and renal (about one third), in rabbits the contribution of the two routes is about equal. Thallium is also secreted in saliva.

As with many other substances, the excretion of thallium in humans differs from that in laboratory animals, since the rate of excretion is generally much lower in humans (rate constant = 0.023-0.069 day^{-1}) than in laboratory animals (average rate constant = 0.18 day^{-1}). Another major difference between humans and animals is the relative contribution of the different routes of excretion. In humans, renal excretion seems to be much more important than in animals, although its relative contribution to the total body clearance has not been definitively established, due principally to the lack of sufficient human data. Moreover, exposure levels, duration of exposure, impairment of excretory organ function, potassium intake and concomitant treatment of acute poisoning may considerably influence the results.

In one study renal excretion of thallium was reported to be about 73%, whereas that through the gastrointestinal tract was about 3.7% of the daily excreted amount. Excretion through hair and skin, and sweat has been estimated to be 19.5% and 3.7%, respectively.

The biological half-life of thallium in laboratory animals generally ranges from 3 to 8 days; in humans it is about 10 days but values up to 30 days have been reported.

No data on the biotransformation of thallium are available.

1.6 Effects on laboratory mammals and *in vitro* test systems

There are no striking species-specific differences in the toxicity of thallium(I) salts. Usually an oral intake of 20 to 60 mg thallium/kg body weight is lethal within one week. Guinea-pigs are slightly more sensitive than other experimental animals. The water-insoluble thallium(III) oxide shows a somewhat lower acute toxicity by oral or parenteral administration than thallium(I) salts. Comparison of acute toxicity data indicates a high degree of bioavailability from all exposure routes. Most organs are affected, but the signs of poisoning and the sequence in which they occur reveal some intra- and interspecies variability.

The symptoms of acute intoxication generally follow the following sequence: firstly anorexia, vomiting and depression, later diarrhoea, skin changes (inflammation at body orifices, skin furuncles, hair loss), and then dyspnoea and nervous disorders. Finally, respiratory failure leads to death.

Symptoms of chronic intoxication are similar to those of acute intoxication. Loss of hair regularly occurs.

Histological examination reveals necrosis or other cell damage. Necrotic changes have been observed in the kidneys, liver, intestine, heart and the nervous system. Swelling of mitochondria and loss of cristae, dilatations of smooth endoplasmic reticulum, increased numbers of autophagic vacuoles and lipofuscin granules, and loss of microvilli have been observed in many cells. The thallium-induced alterations of functional processes may arise from physical disruption of the membranes of subcellular organelles. In the heart, arrhythmogenic effects are restricted to the sinus node.

Thallium intoxication causes selective impairment of certain behavioural elements, which are correlated with biochemical effects (which indicate cellular damage) in certain regions of the brain. Some neurological effects seem to be caused by direct action, e.g. ataxia and tremor by cerebellar alterations or alterations in endocrine activity through changes in the hypothalamus. The autonomic nervous system, mainly the adrenergic, may be activated by thallium. In peripheral nerves, thallium seems to interfere presynaptically, with the spontaneous release of transmitter, by antagonizing these calcium-dependent processes.

The exact mechanism of thallium toxicity is still unknown. Several, perhaps interconnected, mechanisms have been postulated. An important aspect of thallium intoxication is the significant increase in lipid peroxidation and in the activity of the lysosomal enzyme β-galactosidase. The resulting deficiency of glutathione leads to the accumulation of lipid peroxides in the brain and, presumably, finally to lipofuscin granules. The mode of action of thallium seems to be mainly due to a disturbance of the function of the mitochondria.

Sexual activity is usually reduced in chronically poisoned animals, and gonadotoxic effects of thallium are evident in the male reproductive system. In the testes of rats given 10 mg thallium/litre in the drinking-water for 16 days, the Sertoli cells were most sensitive, and desquamation of the spermatogenic epithelium led to immature sperm cells in the semen. This could explain the decreased survival rate of embryos or reduced life span of offspring after sublethal thallium-poisoning of the fathers.

Teratogenic effects, growth inhibition and disturbances in the development of bones were found to occur in chicken embryos after injection of thallium into the egg, but such effects in mammals, even at maternotoxic doses, are controversial. Although transplacental transfer has been demonstrated, many strains of mice and rats show no or only slight teratogenic effects.

Two microbiological mutagenicity tests in *Salmonella typhimurium* were negative and *in vivo* tests on sister chromatid exchange were controversial. However, single studies report chromosomal aberrations or a significant increase of single-stranded DNA breaks.

Long-term studies on the carcinogenicity of thallium are lacking.

1.7 Effects on humans

Since thallium salts are tasteless, odourless, colourless, highly toxic, were easily obtainable in the past and still are in some developing countries, thallium has often been used for suicide, homicide and attempts at illegal abortion, causing acute thallium poisoning. Indeed, thallium intoxication is considered one of the most frequent causes, on a worldwide scale, of purposeful or accidental human poisoning. Knowledge of chronic thallium intoxication is limited to occupational exposure, to population groups in contaminated areas and to cases of homicide involving multiple low doses.

Symptoms of acute thallium toxicity depend on age, route of administration and dose. Doses which have proved lethal vary between 6 and 40 mg/kg, being on average 10 to 15 mg/kg. Without therapy this average dose usually results in death within 10 to 12 days, but death occurring within 8-10 h has also been reported.

The triad of gastroenteritis, polyneuropathy and alopecia is regarded as the classic syndrome of thallium poisoning, but in some cases gastroenteritis and alopecia were not observed. Several other signs and symptoms also occur, varying in order, extent and intensity.

Symptoms of thallium intoxication are often diffuse and initially include anorexia, nausea, vomiting, metallic taste, salivation, retrosternal and abdominal pain and occasionally gastrointestinal haemorrhage (blood in faeces). Later, constipation is commonly seen and may be resistant to treatment, thus interfering with antidotal treatment.

After 2 to 5 days some of the typical thallium disorders slowly develop, irrespective of the route of exposure. Effects on the central and peripheral nervous system vary, but a consistent and characteristic feature of thallium intoxication in humans is the extreme sensitivity of the legs, followed by the "burning feet syndrome" and paraesthesia. Involvement of the central nervous system (CNS) is indicated by symptoms like hallucinations, lethargy, delirium, convulsions and coma. Common circulatory symptoms are hypertension, tachycardia and, in severe cases, cardiac failure. Loss of head hair and sometimes body hair occurs after the second week of poisoning; dystrophy of the nails is manifested by the appearance of white lunular stripes (Mee's lines) 3 to 4 weeks after intoxication. The black regions found in hair papillae are not caused by deposition of pigments or thallium but are due to small amounts of air entering the shaft.

In lethal cases the time until death occurs may vary from hours to several weeks, but most commonly death occurs within 10 to 12 days. Causes of death are mainly renal, CNS and cardiac failure.

In sublethal poisonings, recovery often requires months. Sometimes neurological and mental disturbances as well as electroencephalographic abnormalities and blindness can remain. Additionally, intellectual functions seem to be adversely affected in survivors.

In cases of chronic poisoning, symptoms are similar but in general milder than in cases of acute intoxication. Sometimes permanent blindness occurs. Complete recovery takes months and can be interrupted by relapses.

In a well-investigated case of thallium emission around a cement plant in Lengerich, Germany, thallium concentrations in the hair and urine of exposed people did not correlate with certain features which are known to be usually associated with chronic thallium poisoning, but only with subjective neurological symptoms.

Postmortem examinations or biopsies following thallium poisoning reveal damage of various organs. For example, after ingestion of lethal doses, haemorrhages in the mucosa of the intestine, lung, endocrine glands and heart, fatty infiltrations in liver and heart tissue, and degenerative changes to glomeruli and renal tubules occur. In the brain, fatty degeneration of ganglion cells, damage to axons and disintegration of myelin sheaths can be observed.

Variations in blood pressure may be caused by direct effects of thallium on the autonomic nervous system. Thallium intoxication causes symmetric, mixed peripheral neuropathy. Distal nerves are affected more than proximal nerves, and earlier but lesser degrees of damage occur in nerves with shorter axons, e.g., cranial nerves. Axons are swollen and contain vacuoles and distended mitochondria. In lethal poisoning, severe damage of the vagus nerve, denervation of the carotid sinus and lesions of the sympathetic ganglia have been observed. In sublethal poisoning, affected nerves may undergo axonal degeneration with no or only partial recovery within 2 years.

Retrobulbar neuritis and resulting visual disorders can develop and persist for months after terminating treatment with thallium-containing depilatories, and even optic atrophy may occur.

Limited data are available on the effects of thallium on human reproduction. Menstrual cycle, libido and male potency may be adversely affected. Effects on sperm are known to occur following chronic intoxication. As in animal studies, transplacental transfer occurs; this was seen following a thallium-induced abortion. However, apart from a relatively low weight and alopecia of newborn babies, fetal development was not affected in about 20 cases of thallium intoxication during pregnancy.

No reports of any carcinogenic effects or data on immunological effects of thallium are available. There is no adequate evidence of genotoxic effects.

Therapies of thallium intoxication combine forced diuresis, use of activated charcoal and prevention of re-absorption in the colon by administration of Prussian Blue, potassium ferric hexacyanoferrate(II).

1.8 Human dose-response relationship

The mean urinary thallium concentration in unexposed populations is 0.3 to 0.4 μg/litre. As thallium has a short biological half-life, measured in days, and assuming steady-state conditions, this urinary concentration can be taken as an indicator of total dose following inhalation and dietary intake.

The mean urinary thallium concentration in a population sample living near a thallium atmospheric emission source was 5.2 μg/litre. A clear dose-response relationship was found between urinary thallium concentration and the prevalence of tiredness, weakness, sleep disorders, headache, nervousness, paraesthesia, and muscle and joint pain. A similar dose-response relationship was also reported when thallium in hair was used as an indicator of exposure.

The Task Group considered that exposures causing urinary thallium concentrations below 5 μg/litre are unlikely to cause adverse health effects. In the range of 5-500 μg/litre the magnitude of risk and severity of adverse effects are uncertain, while exposure giving values over 500 μg/litre have been associated with clinical poisoning.

1.9 Effects on other organisms in the laboratory and field

Thallium affects all organisms, but species- and also strain-specific differences are evident. Different inorganic thallium(I) and thallium(III) compounds and organothallium compounds can show different toxicities.

The most important effect of thallium on microorganisms seems to be inhibition of nitrification by soil bacteria. Results of one study suggest that microbial community structure is disturbed at soil concentrations in the range of 1-10 mg/kg dry weight, but the form of thallium used in this experiment was not identified.

Thallium is taken up by all plant parts, but principally by the roots. After uptake into the cell, it is concentrated unevenly in the cytosol, probably bound to a peptide. Thallium concentrations found in plants depend on soil properties (especially pH, clay and organic matter content), as well as on the developmental stage and on the part of the plant. Thallium accumulates in chlorophyll-containing regions, but to a lesser degree in thallium-resistant plants. Oxygen production is reduced by thallium, presumably by direct action on electron transfer in photosystem II. Interference with the pigments is indicated by the occurrence of chlorosis. In addition, impaired uptake of trace elements seems to be involved in the mechanism of toxicity. Growth is also affected, roots reacting more sensitively than leaves or stems. These effects have been reported at concentrations as low as 1 mg thallium/kg of dry plant tissue, after exposure to monovalent forms of thallium.

Most studies of effects on aquatic organisms have used soluble monovalent thallium compounds. The lowest thallium concentration reported to affect aquatic species is 8 μg/litre, which caused a reduction in growth of aquatic plants. Invertebrates are often affected at lower concentration than fish (96-h LC_{50} values are 2.2 mg thallium/litre for daphnids and 120 mg/litre for a freshwater fish). The lowest LC_{50} value, reported after exposure for about 40 days, was 40 μg/litre for fish.

Many cases of thallium intoxication of wildlife have been caused by its large scale application as a rodenticide. In seed-eating animals and predators the CNS and/or the gastrointestinal tract are most severely affected. These effects can also be observed in farm animals. In addition, thallium causes a loss of dorsal feathers in ducks, salivation from the nose and mouth of cattle, and reduced growth in broilers, laying hens, sheep and steers.

2. IDENTITY, PHYSICAL AND CHEMICAL PROPERTIES, AND ANALYTICAL METHODS

Thallium is a soft, malleable, heavy metal with a bluish-white colour and the chemical symbol Tl. The name thallium derives from the beautiful green spectral line (*thallos*, green shoot), which identified the element.

An overview on the properties, synonyms and chemical formulae of pure thallium and some of its compounds is given in Table 1.

2.1 Identity

Thallium is the fifth element in Group IIIB of the Periodic Table. It occurs naturally as two isotopes thallium-203 and thallium-205 with abundancies of 29.52 and 70.467%, respectively (Aderjan et al., in press). The relative atomic mass of thallium is 204.383, the atomic number is 81, and the electron configuration is (Xe) 4f14 5d10 6s2 6p. Due to its high specific gravity of 11.85 g/cm^3, thallium belongs to the heavy metal group, which comprises all metals with a specific gravity of over 4.5 g/cm^3 (Micke et al., 1983).

2.2 Physical and chemical properties

The physical properties of elemental thallium are similar to those of lead; it is very soft and malleable. Thallium exists in both the monovalent (thallous) and the trivalent (thallic) form. Because the 6s electrons possess only a low tendency to be released or bound covalently, the thallous form is more common and stable and forms numerous stable salts. Thallium(III) is easily reduced to thallium(I) by reducing agents at high temperatures (Tl^+/Tl^{3+} = +1.12v) (Micke et al., 1983; Schoer, 1984; Stokinger, 1987).

Below 234 °C the metal crystallizes in a hexagonal close-packed form (α-thallium), while at 234 °C it converts to the β-form, a cubic body-centred lattice. Thallium begins to volatilize at 174 °C. It has a melting point of 303 °C, a boiling point of 1457 °C and a normal potential of Tl/Tl^+ -0.335v (Micke et al., 1983). Thallium is a very reactive metal. When exposed to air and moisture, it is superficially oxidized, forming a coating of thallium(I) oxide (Tl_2O). At higher temperatures it reacts with a lovely green flame to form thallium(III) oxide (Tl_2O_3). Thallium

Table 1. Physical and chemical properties of thallium and some selected thallium compounds[a]

Name	Chemical formula	CAS registry number	Relative atomic/molecular mass	Specific gravity (g/cm^3)	Melting point (°C)	Boiling point (°C)	Colour	Solubility in water (g/litre)
Thallium	Tl	7440-28-0	204.38	11.85	303.5	1457	bluish-white, metallic	insoluble
Thallium(I) acetate	$TlC_2H_3O_2$	563-68-8	263.43	3.765	131	-	silky white	very soluble
Thallium aluminium sulfate	$TlAl(SO_4)_2.12H_2O$	52238-56-9	639.66	2.306	91	-	colourless	117.8
Thallium(I) bromide	TlBr	7789-40-0	284.29	7.557 (17.3 °C)	480	815	pale yellow	0.5 (25 °C)
Thallium(I) carbonate	Tl_2CO_3	29809-42-5	468.78	7.110	273	-	white	40.3 (15.5 °C)
Thallium(I) chloride	TlCl	7791-12-0	239.84	7.004 (30 °C)	430	720	white	2.9 (15.5 °C)

Table 1 (contd).

Name	Chemical formula	CAS registry number	Relative atomic/molecular mass	Specific gravity (g/cm^3)	Melting point (°C)	Boiling point (°C)	Colour	Solubility in water (g/litre)
Thallium(III) trichloride	$TlCl_3$	13453-32-2	310.74	-	25	decomposes	colourless, hygroscopic	very soluble
	$TlCl_3.4H_2O$	13453-33-3	382.80	-	37	100 ($-H_2O$)	colourless	862
Thallium ethylate	$TlOC_2H_5$	20398-06-5	249.44	3.493 (20 °C)	-3	130 (decomposes)	colourless	-
Thallium(I) fluoride	TlF	7789-27-7	223.38	8.23 (4 °C)	327	655	colourless	786 (15 °C)
Thallium(III) trifluoride	TlF_3	7783-57-5	261.38	8.36 (25 °C)	550 (decomposes)	-	olive green	decomposes to TlOH
Thallium hydroxide	TlOH	1310-83-4	221.39	-	139 (decomposes)	-	pale yellow	259
Thallium(I) iodide (α)	TlI	7790-30-9	331.29	7.29	440 (β)	823 (β)	yellow	0.006 (20 °C)

Table 1 (contd).

Name	Chemical formula	CAS registry number	Relative atomic/molecular mass	Specific gravity (g/cm^3)	Melting point (°C)	Boiling point (°C)	Colour	Solubility in water (g/litre)
Thallium(I) nitrate (α)	$TlNO_3$	10102-45-1	266.39	-	206	430	white	95.5 (20 °C)
Thallium (III) nitrate trihydrate	$Tl(NO_3)_3 \cdot 3H_2O$	13453-38-8	444.44	-	105-107	decomposes	colourless	decomposes
Thallium(I) oxide	Tl_2O	1314-12-1	424.77	9.52 (16 °C)	300	1080 (-O)	black	decomposes to TlOH
Thallium(III) oxide	Tl_2O_3	1314-32-5	456.76	10.19 (22 °C)	717 ± 5	875 ($-O_2$)	black	insoluble
Thallium(I) sulfate	Tl_2SO_4	7446-18-6	504.82	6.77	632	decomposes	white	48.7 (20 °C)
Thallium(I) sulfide	Tl_2S	1314-97-2	440.85	8.46	448.5			0.2 (20 °C)

[a] From: Stokinger (1987); Budavari (1989); Lide (1990)

carbonate (Tl_2CO_3) is the only heavy metal carbonate that is very soluble in water (Micke et al., 1983; Stokinger, 1987).

In contact with water, thallium(I) hydroxide is formed from the metal. Thallium is very soluble in HNO_3 and H_2SO_4, but only slow dissolution takes place in HCl, because of the low solubility of the halides. It is insoluble in alkali bases. Thallium combines with fluorine, chlorine and bromine at room temperature, and reacts with iodine, sulfur, phosphorus, selenium and tellurium after heating. The metal does not react with molecular hydrogen, nitrogen or carbon. It forms alloys with other metals and readily amalgamates with mercury (Micke et al., 1983).

The ionic radii and the electronegativity constant of monovalent thallium are very similar to those of other alkali metals. Thallium(I) hydroxide, carbonate and sulfate, like the corresponding potassium compounds are very soluble in water. With respect to their physical and chemical properties, e.g., poor water solubility, thallium(I) oxide, sulfide and halides show similarities to the corresponding compounds of silver, mercury and lead (Trotman-Dickenson, 1973). In contrast to inorganic thallium compounds, covalent organothallium compounds are only stable in the trivalent form (McKillop & Taylor, 1973). Thallium(I) is not strongly complexed by humic acids, whereas thallium(III) forms stable complexes of the $[TlX_4]^-$ or $[TlX_6]^{3-}$ type (Schoer, 1984).

2.3 Conversion factor

1 g thallium = 0.0049 mol
1 mol thallium = 204.38 g

2.4 Analytical methods

Classical analytical methods, the introduction of new techniques and a combination of both with enrichment or separation processes provide suitable methods for the quantitative detection of thallium in various media. Because thallium concentrations in environmental samples are very low, determination directly from the sample or from the digestion solution usually lacks sufficient accuracy. Therefore, preconcentration procedures are necessary (Schoer, 1984; Sager & Tölg, 1984).

2.4.1 Sampling and sample preparation

Thallium losses during sampling, sample preparation and determination are a major source of analytical error.

Contamination hazards need to be anticipated, as thallium is present in laboratory ware and is leached out by solutions (Kosta, 1982). Glass contains about 1-10 μg thallium/kg. Leaching of polythene containers with 6M HCl for 1 week brought 1-10 ng thallium/cm^2 into solution. In addition, thallium(I) in 0.1M HNO_3 solution adsorbs onto container walls made of polyethene, polypropene, glassware or rubber. This effect depends on the chemical properties of the surface of the container walls and on the concentration of matrix ions. At a thallium concentration of 1 mg/litre, no losses to borosilicate surfaces at $pH < 4$ were reported, but extensive adsorption occurred at $pH > 10$ (Sager & Tölg, 1984).

For determinations with spectrophotometric, mass spectrometric, voltametric and other methods, digested samples are needed. With respect to the high volatility of the metal and the low boiling points of some of its compounds, only closed systems are recommended for the digestion of organic matrices to prevent thallium losses. Fusion, dry ashing and fuming with HF and H_2SO_4 or $HClO_4$ may lead to severe losses (up to 40%) of the thallium present (Matthews & Riley, 1969). High-pressure digestion in closed quartz vessels with concentrated acids, e.g., HNO_3 or HNO_3 and HF, at temperatures up to 300 °C is the most suitable procedure for nearly all matrices (Knapp, 1985). HF interferes with analysis by GFAAS or ICP-AES and needs to be removed by heating to dryness with H_3BO_3 (Han et al., 1982).

The volatility of thallium and its oxide or chloride makes it possible to separate these with a gas stream of O_2, H_2 or HCl from other elements that do not form volatile components under the same conditions and subsequently capture them in a cool trap. This procedure can be used as a preconcentration step when large quantities of sample are available (Geilmann & Neeb, 1959; Han et al., 1982; Sager, 1984).

Other methods of preconcentration are coprecipitation, anodic electrolysis, ion exchange and liquid-liquid extraction. Coprecipitation is not selective, but it leads to a high concentration factor and results in a definite matrix, which might be useful in some methods (Griepink et al., 1988). For example, coprecipitation with $Fe(OH)_3$ leads to separation from a salt matrix (K^+, NH_4^+). Electrolytic deposition or cementation with zinc powder yields an excellent separation, although this procedure is time-consuming. Ion exchange, which gives a specific separation in certain cases, is also time-consuming. Liquid-liquid extraction with chelating agents is virtually nonspecific, but it is a fast and easy method. A disadvantage is the relatively low concentration factor (Sager & Tölg, 1984).

Isotope dilution methods have been applied to avoid ionization matrix effects. Thallium is measured as thallium-205; the thallium-203 isotope can be used as a spike for isotope dilution (Sager, 1986).

2.4.2 Methods of determination

Thallium is almost always determined as total metal, rather than as specific thallium compounds. Among the analytical techniques that can be used are spectrophotometry, mass spectrophotometry (MS), atomic absorption spectrometry (AAS), voltametry, neutron activation analysis (NAA), X-ray fluorimetry, and inductively coupled plasma (ICP) techniques (Sharma et al., 1986). A selection of analytical methods is summarized in Tables 2 and 3.

2.4.2.1 Atomic absorption spectrometry

The most widely used method of thallium determination is atomic absorption analysis, using measurement at 276.8 nm with a thallium hollow cathode lamp. The sensitivity can be improved by the use of an electrode discharge lamp (EDL), owing to its higher intensity. Graphite furnace atomic absorption spectrometry (GFAAS) is a well-established technique for the monitoring of trace elements in nearly all kinds of matrices. The technique has sufficiently low detection limits and is well-suited to applications where high sensitivity is required for small sample amounts. In Table 3 some methods for GFAAS are summarized.

The platform furnace concept in the temperature-stabilized mode, together with Zeeman effect background correction, allows almost interference-free determinations of many elements. Sample pretreatment is not necessary, which greatly reduces the risk of substance losses or contamination of the sample prior to analysis (Minoia et al., 1990).

Matrix modifiers permit higher pyrolysis temperatures, so that the desired element can be isolated from matrix elements and compounds in an ideal case. Letourneau et al. (1987) found that additions of H_2SO_4 as a matrix modifier were inadequate and that interferences could not be corrected by Zeeman background

Table 2. Instrumental methods for the determination of thallium

Method[a]	Matrix	Oxidation state	Sample pretreatment	Parameters of method	Interferences	Detection limit	Reference
PAA	metals	-	^{203}Tl (γ,n) ^{202}Tl 30 MeV bremsstrahlung post-irradiation separation of Tl from the matrix	γ440 keV			Segebade & Schmitt (1987)
NAA	biological material	post-irradiation extraction	drying, 2-M	10^{13} n/cm^2.sec 3-7 days ^{203}Tl (n,γ) ^{204}Tl 0.77 MeV β^- measurement	other isotopes than ^{204}Tl	1 μg absolute	Itawi & Turel (1987)
AMP	water	Tl (I)	Na_2CO_3, $NaHCO_3$, thiomalic acid	-0.47 vs sat. calomel electrode	Mn(VII), Co(II), Sn(II), Tl(III)	-	Agrawal & Khatkar (1988)
DPASV	urine, saliva	-	Na acetate, $HClO_4$, EDTA	-1.0 vs sat. calomel electrode	Cd, Pb	0.2 μg/litre	Vandenbalck & Patriarche (1987)
ICP-MS	rocks	-	HNO_3, HF, H_2O_2	-	polyatomic interferences	70 ng/litre	Date et al. (1988)

Table 2 (contd).

GDMS	indium	-	-	at pressure 3.10^{-4} mbar discharge voltage 1 kV discharge current 3 mA accelerating voltage 8 kV resolution 4000	-	30 μg/kg	Guidoboni & Leipziger (1988)
ICP-AES	air particulates	-	$HNO_3/HClO_4$ (4:1)	190.9 nm	F^-	17 μg/litre leachate	NIOSH (1984)
ICP-AES	biological material	-	Parr bomb		F^-	0.05-0.1 mg/litre	Que Hee & Boyle (1988)
ICP-MS	water	-	HNO_3	^{205}Tl	-	0.1 μg/litre	Henshaw et al. (1989)
ICP-MS	tissues	-	HNO_3	-	-	18 μg/kg	Templeton et al. (1989)
Spectrometry	environmental	Tl (I)	dithizone, $CHCl_3$, EDTA, citrate, cyanide	-	Ag, Hg	1 μg/litre	Sager (1986)

[a] AMP = amperometric titration; DPASV = differential pulse anodic stripping voltametry; GDMS = glow discharge mass spectrometry; ICP-AES = inductively coupled plasma - atomic emission spectrometry; ICP-MS = inductively coupled plasma - mass spectrometry; NAA = neutron activation analysis; PAA = photon activation analysis; PPS = proton-induced prompt low energy photon high resolution spectrometry

Table 3. Methods for determining thallium (Tl) with graphite furnace atomic absorption spectrometry (GFAAS)

Sample	Separation	Injected solution	Detection limit	Interferences	Reference
Fly ash, soil	digestion and preconcentration by extraction of Tl(III) with diiso-propylether evaporation	diluted H_2SO_4, HNO_3	3.3 ng/litre (including the 400 x preconcen-tration step)	HBr	De Ruck et al. (1989)
Urine	complex with tri-*n*-octylamine, extraction with ethanol into 5 mg metallic *n*-butyl acetate gallium	organic layer diluted with ethanol and H_2SO_4	0.3 μg/litre	max. charring temp. 400 °C	Flanjak & Hodda (1988)
Gallium	-	gallium	200 μg/kg	-	Hiltenkamp & Jackwerth (1988)
Urine	-	spiked urine, diluted	2 μg/litre	NaCl	Berndt & Sopczak (1987)
Urine	chelation with NaDDC, extraction with MIBK	MIBK extract	0.05 μg/litre	-	Apostoli et al. (1988)

Table 3 (contd).

Sample	Separation	Injected solution	Detection limit	Interferences	Reference
Mineralized faeces and tissues	-	HNO_3, H_2SO_4, ascorbic acid, Triton X-100	5 µg/litre	NaCl	Leloux et al. (1987a)
Blood, serum	-	HNO_3	10 µg/litre		Leloux et al. (1987a)
Erythrocytes	-	HNO_3	12 µg/litre		Leloux et al. (1987a)
Soil, sediments	extraction with tri-octyl-methylammonium, MIBK		20 µg/kg	Cu, Zn, Pb	Ebarvia et al. (1988)
Coal fly ash	-	HNO_3	-	-	Bettinelli et al. (1988)

MIBK = methyl isobutyl ketone; NaDDC = sodium diethyl dithiocarbamate compensation. Modifying the matrix with palladium and magnesium nitrate has been suggested to be generally applicable, but this is not as effective for thallium as it is for other elements (Welz et al., 1988a). A combination of 6 mg palladium with 100 mg ammonium nitrate allows the direct determination of thallium in ten-fold diluted blood against matrix-free standards (Yang & Smeyers-Verbeke, 1991).

Paschal & Bailey (1986) determined thallium concentrations in urine. The samples were diluted 1:1 with a matrix modifier consisting of magnesium nitrate, HNO_3, Triton X-100 and water. The detection limit was calculated to be 0.5 μg/litre.

Chemical interferences due to chloride ions are important. These interferences are caused by volatilization of thallium chloride in the pyrolysis stage and, in part, by formation of TlCl(g) during the atomization stage. Even matrix modification gives unsatisfactory results. Welz et al. (1988b) showed that addition of palladium nitrate as a modifier and application of argon with 5% H_2 as a purge gas leads to interference-free determination with, for instance, NaCl loads of up to 100 mg. A special pre-pyrolysis step is necessary to reduce palladium to the metal state, thus enabling adsorbed H_2 to react with the chloride compounds to form volatile HCl. Similar results were obtained by Manning & Slavin (1988).

De Ruck et al. (1987) reported an oxidation technique for natural waters with cerium(IV) sulfate and a subsequent preconcentration step on an anion-exchange column. A preconcentration factor of 400 was achieved, and the resultant detection limit was 3.3 ng/litre using GFAAS. Flame atomic absorption is a reliable method for measurement of thallium concentrations at the level of mg/litre or more. The determination is easy and free from interference (Welz, 1983; Griepink et al., 1988).

2.4.2.2 Inductively coupled plasma - mass spectrometry

ICP-MS is a promising method for concentrations in the μg/kg range or less, and has good precision and accuracy. It is a multi-element technique with sub-ppb detection limits for many elements. Additional advantages of mass discrimination include its suitability for isotope ratio analysis and stable isotope tracer analysis, and the extended range of elements that can be studied. Some ICP-MS methods are summarized in Table 2.

The application of ICP-MS to the analysis of thallium in iron-rich ores was described by Date et al. (1988). No polyatomic interferences for iron were detected in acid solutions. The addition of 500 mg iron/litre to a solution of 1 mg thallium/litre in 1% HNO_3 resulted in a 0.1% increase in the thallium peak. The detection limit was found to be 0.07 μg/litre.

Templeton et al. (1989) examined thallium concentrations in rat liver and blood plasma samples which were submitted to acid digestion and reported a detection limit of 0.09 μmol/kg (18 μg/kg).

More than 250 water samples from lakes were analysed for thallium (thallium-205) by ICP-MS after acidification with HNO_3. The detection limit was found to be 0.1 μg/litre; the recovery of spiked analytes amounted to 112 ± 4% (Henshaw et al., 1989).

2.4.2.3 *Other methods*

Methods other than AAS and ICP-MS are summarized in Table 2. Spectrophotometric determination with rhodamin B after liquid/liquid extraction is a quick and easy method, but it is less sensitive and has a high incidence of interference. The method is suitable for a quick visual test, when a massive intoxication with thallium compounds is suspected. Determinations down to 10 μg/litre are possible in environmental matrices (Griepink et al., 1988).

Inductively coupled plasma - atomic emission spectrometry (ICP-AES) is a rapid multi-element technique, but it does not provide the detection limits required to measure thallium concentration in uncontaminated samples. The NIOSH method for determining thallium in air particulates has a detection limit of 17 μg/litre of leaching solution (NIOSH, 1984).

Differential pulse anodic stripping voltametry (DPASV) is a sensitive method for the quantitative determination of thallium in water samples or urine. Voltametric methods also offer the advantage of simultaneous determination of several metals from one solution. The lower limit of detection for thallium(I) is 10-100 ng/litre (Klahre et al., 1978; Vandenbalck & Patriarche, 1987; Griepink et al., 1988).

Neutron activation analysis (NAA) is applicable for the determination of thallium in various environmental samples, but it is relatively slow and impractical for the routine analysis of large numbers of samples. The detection limit is determined by the irradiation time, neutron flux, the choice of a radiochemical separation of the radio-isotope to remove interfering matrix radio-isotopes and the measurement time. Levels down to the absolute amount of ng of thallium can be determined (Schoer, 1984). This method can therefore be used for the determination of low thallium concentrations in biological samples. In bovine liver a detection limit of 1.5 μg/kg was found after digestion, separation and concentration procedures (Henke, 1991).

Thallium(I)-sensitive electrodes are not sensitive enough for trace determinations, and high concentrations of alkali ions reduce the selectivity. Sensitivity problems must also be considered for the usual X-ray fluorimetry techniques. Other methods, like excitation with charged particles and photon activation radiochemical isotope dilution, are seldom used.

2.4.3 *Quality control and quality assurance*

Sample collection, analysis and data presentation should be carried out according to a protocol which ensures adequate validation of biological monitoring procedures (Vesterberg et al., 1993).

There is an urgent need for strict quality control and quality assurance of the analytical data on thallium in clinical and environmental samples. It is only when proof is given for the accuracy of the published data that they become unequivocally useful to establish critical concentrations and dose-response relationships in a given population or ecosystem. General considerations of quality control and quality assurance have been recommended by WHO (WHO, 1986; Aitio, 1988).

To date, very few of the many studies on thallium have provided the necessary evidence concerning the quality of the data throughout the analytical procedure. The recognized way to control and ensure this involves good laboratory practice (GLP), including intra- and inter-laboratory analysis of materials with certified concentrations of thallium. Such Certified Reference Materials (CRMs) should have the same (or a similar) matrix as the sample to be analysed and be certified for thallium concentrations (similar to those in the sample) by an internationally recognized body. This implies suitable levels for thallium in serum, whole blood, urine, faeces, animal tissues and plants, as well as levels typical for exposed individuals, animal studies or eco-systems (Cornelis, 1988).

Available reference materials with clinical and environmental interest are listed in Table 4. This immediately reveals the very poor picture for CRMs certified for thallium. Whole blood and serum samples are totally lacking, while urine of exposed individuals is handled by the BI CUM 2 and 3 products with assigned values for thallium only. The BCR milk powders and the NBS liver samples carry a reference value. Thallium has also been reported in some environmental samples (fly ash, etc.) without being certified.

There appears to have been only one inter-laboratory survey on thallium in two spiked urine samples (Geldmacher-von Malinckrodt et al., 1984). The 35 participating laboratories used one of the three routine methods, AAS, DPASV or photometry, after thallium extraction. The samples were also analysed by IDMS (isotope dilution mass spectrometry) and attributed reference values of 66.3 and 483 μg thallium/litre, respectively. The evaluation of this inter-laboratory survey revealed that about 70% of the laboratories met the goal.

2.4.4 Conclusions

There are several methods available for the determination of thallium in biological and environmental samples. As routine methods these are GFAAS (the most widely used), DPASV, ICP-MS and photometry. They all require a very careful sample pretreatment and, in the case of DPASV and photometry, perfect mineralization of the sample without losses due to volatilization or adsorption onto the container walls. The same remarks apply to the methods including a preconcentration step. In the case of GFAAS and ICP-MS, direct analysis of the diluted sample is feasible. It is strongly recommended that all analyses be accompanied by a quality assurance programme. At present, it is possible to determine thallium concentrations of about 0.1 μg/litre or 0.1 μg/kg.

Table 4. Reference materials for thallium determinations in biological and environmental materials[a]

Matrix	Origin[b]	Code	Thallium concentration	Remarks
Liver	NBS	SRM 1577	50 μg/kg	lyophilized bovine liver
		SRM 1577A	3 μg/kg	lyophilized bovine liver
Milk powder	BCR	CRM 63	1.3 μg/kg	natural skim milk powder
		CRM 150	1.0 μg/kg	spiked milk powder
		CRM 151	0.8 μg/kg	spiked milk powder
Urine	BI	CUM 2	93 ± 13 μg/litre[c]	lyophilized synthetic urine
		CUM 3	603 ± 78 μg/litre[c]	lyophilized synthetic urine
City waste	BCR	BCR-CRM-176	2850 μg/kg	certified; error 6.7%
Coal fly ash	IRANT	IRANT-ECO	14 000 μg/kg	not certified
Coal fly ash	NIST	NIST-SRM-1633a	5700 μg/kg	certified; error 3.5%
Gas coal	BCR	BCR-CRM-180	2200 μg/kg	not certified
Steel plant flue dust	IRANT	IRANT-OK	< 3000 μg/kg	not certified

[a] According to Muramatsu & Parr (1985) and Cortes Toro et al. (1990)

[b] BCR: Measurement and Testing Programme, DG XII, BCR, Commission of the European Union, Wetstraat 200, B-1049 Brussels, Belgium
BI: Behring Institute, PO box 140, D-3350 Marburg 1, Germany
IRANT: Institute of Radioecology and Applied Nuclear Techniques (CSSR)
NBS (new name NIST): Room B 311, Chemistry Building, National Institute for Standardization and Testing, Gaithersburg, MD 20899, USA
NIST: National Bureau of Standards (USA)

[c] assigned values for a particular lot only

3. SOURCES OF HUMAN AND ENVIRONMENTAL EXPOSURE

3.1 Natural occurrence

Thallium is ubiquitous in nature, but occurs at low concentrations (< 2 mg/kg) (section 5.1), especially in sulfide ores of various heavy metals (zinc, copper, iron and lead) and in minerals of potassium, caesium and rubidium (Micke et al., 1983; Kemper & Bertram, 1984; Ohnesorge, 1985; Stokinger, 1987; Manzo & Sabbioni, 1988). Although the concentration of thallium is low, about 700 000 tonnes of thallium are contained in worldwide identified resources of coal and 19 000 tonnes in zinc resources (US BM, 1989). There are only a few areas with a naturally very high thallium concentration, e.g., the Alsar in the Former Yugoslav Republic of Macedonia (Zyka, 1972). Minerals of thallium, e.g., lorandite ($TlAsS_2$) and crookesite ($(Cu,Ag,Tl)_2Se$), with thallium concentrations of up to 60%, are rarely found and usually not used for production of thallium (Micke et al., 1983; Kemper & Bertram, 1984; Briese et al., 1985; Kazantzis, 1986).

3.2 Anthropogenic sources

3.2.1 *Production levels and processes*

Since thallium is used only in small amounts by industry, worldwide production of pure thallium is low. In 1975 about 8 tonnes were produced in Germany and 2 to 3 tonnes in the USA (Zitko, 1975a), while in 1987 and 1988 worldwide production was about 17 tonnes (US BM, 1992). In 1981 the production of thallium in the USA was discontinued. Sources for the production of thallium are zinc, lead and sometimes copper or iron smelters and sulfuric acid plants. Flue dust in particular is used as a thallium source (Zitko, 1975a; Smith & Carson, 1977; Micke et al., 1983; Briese et al., 1985). Procedures for the separation of thallium from other metals depend on the proportions of the different minerals and, therefore, vary considerably between the different smelters (Sanderson, 1952; Smith & Carson, 1977; Micke et al., 1983; Kemper & Bertram, 1984; Briese et al., 1985).

3.2.2 *Uses*

Thallium(I) sulfate was once used in medicine to reduce sweating and to cure various infections, e.g., venereal diseases, ringworm of the scalp, typhus, tuberculosis and malaria, and as a depilatory agent, which caused many intoxications (Munch, 1934b; Smith & Carson, 1977; Emsley, 1978; Briese & Nessler, 1985a). However, therapeutic uses of thallium have been discontinued because of its toxicity. Since 1920, thallium(I) sulfate has been used as a rodenticide, in Europe chiefly against rats and in the USA chiefly against ground squirrels (Howe, 1971; Smith & Carson, 1977). Formerly it was used as an insecticide (against ants and cockroaches). However, thallium is no longer on sale as a rodenticide in most industrial countries (Bruère et al., 1990), but is still used in developing countries because of its cheapness.

Other areas in which thallium is used (Howe, 1971; Smith & Carson, 1977; Micke et al., 1983; Briese et al., 1985; Sharma et al., 1986; Kazantzis, 1986; Manzo & Sabbioni, 1988; ATSDR, 1992) are as follows:

a) low temperature thermometers (down to -59 °C) made from a mixture of mercury and thallium;

b) special glasses with a high resistance and a low melting point, containing thallium and selenium;

c) mixed crystals for infrared instruments, composed of arsenic or thallium(I) salts and halogenides (TlI-TlBr), and Tl_3VS_4, Tl_3NbS_4, and Tl_3PSe_4 for acusto-optic and laser equipment;

d) electronic devices, e.g. thallium(I) sulfide for semiconductors and scintillation counters;

e) mercury lamps (addition of thallium(I) halogenids increases the yield of light and changes its spectrum);

f) alloys with lead, zinc, silver and antimony enhance resistance to corrosion;

g) catalysing organic reactions, e.g., oxidations of hydrocarbons and olefins (thallium compounds are being increasingly used for organic synthesis); patents summarized by Smith & Carson (1977);

h) radioactive isotopes, used in physics for measurement of exact time periods (thallium-205), in industry for measuring the thickness of material (thallium-204), and in medicine for scintigraphy of heart, liver, thyroid and testes, and for the diagnosis of melanoma (thallium-201) (Rao et al., 1983; Müller-Brand et al., 1984; Urbain et al., 1986);

i) other uses, e.g., in the production of imitation jewellery, fireworks, pigments and dyes, the impregnation of wood and leather against bacteria and fungi, and in mineralogical analysis;

j) minor amounts of thallium are used in biochemistry, e.g., arylthallium(III) compounds for modification of proteins and tRNA (Douglas et al., 1990).

Worldwide industrial consumption in 1991 was estimated to be 10 to 15 tonnes. Between 1940 and 1980 consumption in the USA varied considerably between 0.5 and 11 tonnes/year (Schoer, 1984), and between 1984 and 1988 it was 1.1-1.5 tonnes/year (US BM, 1985, 1989). In the USA it is used mainly in the electrical and electronic industries and the 650 kg used in 1983 in the German Democratic Republic was mainly for making special glass (Smith & Carson, 1977; Micke et al., 1983; Briese et al., 1985; Kazantzis, 1986; Kemper & Bertram, 1991).

3.2.3 *Emissions from industrial sources*

There is an enormous difference between the amount of thallium mobilized (released into air, water or disposed of on land) and the thallium consumption of 12 tonnes/year (section 3.2.2). Worldwide a total of 2000-5000 tonnes of thallium is estimated to be mobilized per year, especially through the combustion of fossil fuels, refinement of oil fractions, the smelting of ferrous and non-ferrous ores, and also by some other industrial processes such as cement production (Gorbauch et al., 1984; Ewers, 1988; Nriagu & Pacyna, 1988). Smith & Carson (1977) estimated that about 15% (240 tonnes) of total mobilized thallium is transferred annually to the atmosphere. However, only a small fraction is released into the atmosphere or wastewater during production processes or from waste materials (Table 5). Summarizing estimations for the USA by Smith & Carson (1977), Schoer (1984) emphasized that in the USA each year nearly 1000 tonnes of thallium are released into the environment, of which 350 tonnes are emitted in vapours and dusts, 60 tonnes bound to non-ferrous metals, and more than 500 tonnes contained in fluid and solid wastes.

Table 5. Estimated emissions of thallium (tonnes/year) into the environment

Emission source	USA	Canada	Germany	Europe	World
Coal combustion					
into air	180[a]		7.5[b]	7[c]	54[d]
	140[e]		4[f]	80[e]	600[e]
			6[g]		
into soil/water	170[a]				
into total environment				240[c]	
Coal combustion (into air)					
from electric utilities					155-620[h]
from industry and domestic					495-990[h]
Ferroalloy production using manganese ores					
into air	140[a]				
into soil/water	220[a]				
Raw iron production and related coal combustion					
into air	6[a]		35[g]		30[d]
into total environment	205[a]				
Production of nonferrous metals					
into air	38[a]	44[i]	11[g]		
total emission	496[a]				
Potash-derived fertilizers					
into total environment	5[a]				
Cement plants					
into air			25[g]		2670-5340[h]
Brick works			28[b]		
Oil fuel combustion, mining and processing of oil shales					
into soil/water	8[a]				
total emission	8[a]				
Waste combustion			< 1[g]		

[a] Smith & Carson (1977)
[b] Brumsack (1977)
[c] Sabbioni et al. (1984b)
[d] Bowen (1979)
[e] Schoer (1984)
[f] Brumsack et al. (1984)
[g] Davids et al. (1980)
[h] Nriagu & Pacyna (1988)
[i] Kogan (1970)

3.2.3.1 Metal production industries

It has been estimated that worldwide over 600 tonnes of thallium are processed per year during the smelting of lead, copper and zinc ores (Micke et al., 1983). Thallium emissions from smelters can vary greatly from plant to plant, depending upon the thallium content of the raw materials and the technology used. For this reason, and because of the lack of recent emission data, global releases can be only roughly quantified. On the basis of the data in Table 5, a total of about 90 tonnes of thallium may be released each year into the atmosphere from non-ferrous metal production operations in the USA, Canada and Germany. Dust in one zinc smelter was reported to contain 380-3700 mg thallium/kg before and 60-9700 mg/kg after starting the production of thallium (Briese et al., 1985). Although it is not possible to estimate the losses of thallium from mineral waste materials, releases from these materials are generally expected to be small.

Data from the USA (Smith & Carson, 1977) indicate that relatively large amounts of thallium are present in waste materials from non-ferrous metal (mainly copper) and iron and steel production (Table 5). Although no precise data were available on thallium levels in waste from ferroalloy production using manganese ores, Smith & Carson (1977) suggested that emissions from this source could be significant. Atmospheric releases resulting from the production of iron and steel in the USA were estimated to be relatively small (about 5 tonnes from steelmaking and 1 tonne in iron blastfurnace gases). In the main area of iron and steel production in Germany, annual thallium emissions into air have been estimated to be about 0.8 tonnes (Ewers, 1988).

3.2.3.2 Power-generating plants

Power-generating plants represent a major source of thallium emissions, especially those using some brown coal or coal of the Jurassic period. Most coals contain only about 0.5 to 3 mg/kg, mainly incorporated in sulfide inclusions. Some of these impurities can be removed by washing and mechanical cleaning. It has been estimated that about half of the thallium content of coal is emitted into the atmosphere and represents the biggest anthropogenic source (Smith & Carson, 1977) (Table 5). In such estimations, losses from collected fly ash are not taken into consideration, because its use may vary. Only a minor amount is used in cement making. If it is used as a soil stabilizer, contamination of the environment is much higher (Smith & Carson, 1977).

Natusch et al. (1974) found that coal-fired power-generating plants emitted about 700 μg thallium/m^3 flue gases, resulting in a local level of air emission of about 700 ng/m^3. This would result in an estimated daily absorbed amount of 4.9 μg airborne thallium per person (US EPA, 1980). In the European Union, coal-fired power-generating plants were estimated to have caused a total mobilization of 240 tonnes of thallium during 1990, about one third of this being concentrated in the smallest particles, and atmospheric emissions of 7 tonnes (Sabbioni et al., 1984b).

In coal burners, thallium volatilizes and recondenses onto the surface of ash particles in cooler parts of the system. As a result, 2 to 10 times higher concentrations of thallium may occur in the fly-ash than was present in the coal (Galba, 1982). Fly-ash thallium content is negatively correlated with particle size (Manzo & Sabbioni, 1988). Thus, thallium and other toxic trace elements are concentrated in the smallest particles, which pass through conventional power-generating plant filters, remain suspended in the atmosphere for long periods and are respirable. For instance, particles with a diameter of 1.1-2.1 μm contain 76 mg thallium/kg fly-ash, those with a diameter of 2.1-7.3 μm contain 62-67 mg/kg and those with a diameter of 7.3-11.3 and > 11.3 μm contain 40 and 29 mg/kg, respectively. Particles with a diameter of less than 74 μm contain only 7 mg thallium/kg (Natusch et al., 1974). These particles are highly toxic, since thallium and other heavy metals are preferentially concentrated on the particle surfaces and therefore are relatively bioavailable (Linton et al., 1976; Natusch, 1982).

3.2.3.3 *Brickworks and cement plants*

Total thallium emissions from brickworks in Germany have been estimated to be 28 tonnes/year. This compares with emissions of 7.5 tonnes/year from the burning of coal (Brumsack, 1977).

The emission potential of cement plants was not recognized until 1979. The first effects on vegetation around a cement plant in Lengerich, Germany were observed in 1977 (Pielow, 1979; LIS, 1980), but only the gradual hair-loss in a rabbit led to the suspicion that thallium was the cause of the toxic effects (LIS, 1980; Brockhaus et al., 1981b; Dolgner et al., 1983). The source of thallium was found to be residues of pyrite roasting added as a ferric oxide additive to powdered limestone in order to produce special qualities of cement and the addition of the filter fly-dust (LIS, 1980). Studies at other plants showed much lower emission levels, so that the emission at Lengerich was caused by the exceptional circumstances. Production alterations in Lengerich caused a reduction in the emissions of more than 99% (Pielow, 1979; Prinz et al., 1979; LIS, 1980).

Like power-generating plants, cement plants emit thallium mainly bound to particles with a diameter of 0.2-0.8 μm (LIS, 1980). Thallium concentrations in fly-dust emitted by the cement plant in Lengerich were about 2.5 mg/m^3 air, of which nearly all was water-soluble thallium(I) chloride. Whereas the filter efficiency was 99% with respect to cement dust, it was only 50% with respect to the thallium-containing particles. As a result, about 140 to 200 g thallium/hour was emitted (Pielow, 1979; Prinz et al., 1979; Weisweiler et al., 1985). Changing the production process reduced the thallium content to less than 25 μg/m^3 (< 200 mg/kg dust). In other cement plants the concentrations in the filter dust were reduced from 3066 mg/kg to about 100 mg/kg, and after this reduction only 13% of the thallium was soluble in water (LIS, 1980).

3.2.3.4 *Sulfuric acid plants*

The sulfuric acid plant which had been the source of the roasted pyrite used in the cement plant in Lengerich used pyrite containing about 400 mg thallium/kg. However, in the roasted pyrite about 7% of the thallium was water-soluble. During production of sulfuric acid, a 100-fold enrichment of thallium was found (LIS, 1980). As a consequence, increased levels of thallium were found in Duisburg, Germany around the sulfuric acid plant but never such high concentrations as around the cement plant (Gubernator et al., 1979).

4. ENVIRONMENTAL TRANSPORT, DISTRIBUTION AND TRANSFORMATION

4.1 Transport and distribution between media

4.1.1 Transport and distribution in air, water and soil

Near point sources such as coal-fired power-generating stations, cement plants using pyrite and some metal smelting operations, the major source of thallium in air is emission of fly-ash (section 5.1.1). Although data on the forms of thallium in these emissions are limited, results of one study indicate that nearly all of the thallium in fly dust from a cement plant (in Lengerich, Germany) was present as soluble thallium(I) chloride (LIS, 1980). No data on the amounts or forms of thallium transported from soil into air during the dry season were identified.

Assuming that 4×10^{12} kg of crustal rocks weather each year, Bowen (1979) calculated that 2.4×10^{6} kg thallium/year become available to plants. However, as Smith & Carson (1977) have noted, thallium tends to be retained during rock weathering, and concentrations in soils tend to be somewhat enriched in soils compared to the original bedrock.

The fate of thallium added to soil in deposited fly-ash probably depends greatly on soil type (Crössmann, 1984). Data from Smith & Carson (1977) suggest that retention should be greatest in soils that contain large amounts of clay, organic matter and iron/manganese oxides. According to McCool (1933) significant amounts of thallium can be removed from solution in soils by ion exchange. Thallium can also be incorporated into stable humus complexes (Crössmann, 1984), which are resistant to rapid "wash-out" (Schoer, 1984).

Results of studies in several areas indicate that thallium deposited from the atmosphere tends to accumulate in the surface layers of soils (Smith & Carson, 1977; Heinrichs & Mayer, 1977; LIS, 1980). For example, after prolonged emissions from a cement plant in Germany (LIS, 1980), thallium was found to remain in the upper levels of soil (Schoer, 1984); material from depths 0-10, 40-50 and 60-70 cm contained 4.2, 1.3 and 0.1 mg thallium/kg, respectively (Schoer, 1984). Retention processes will, however, be less effective in acidic soil. For example, results of studies by

Heinrichs & Mayer (1977) indicate that about 40% of the thallium deposited from the atmosphere onto relatively uncontaminated acidic (pH = 3.9-4.3) forest soil in Germany was leached from the top 50 cm to lower soil horizons. Elevated concentrations of thallium in groundwater (up to 40 μg/litre) and in an irrigation canal (up to 96 μg/litre) in China, near a site where waste materials from the mining of mercuric ore and coal containing 25 to 106 mg thallium/kg were deposited (Zhou & Liu, 1985), indicate that under some circumstances significant amounts of thallium can move from soil into local water.

Although there is little information on the forms of thallium in natural water, most dissolved thallium in fresh water is expected to be present as the monovalent Tl^{+} ion (Smith & Carson, 1977). In strongly oxidizing fresh water and in most seawater (Sager & Tölg, 1984), however, trivalent thallium is probably the predominant dissolved form. Both forms of thallium can be removed from solution by exchange and complexing reactions with suspended solid phases. Trivalent thallium is also susceptible to reduction and precipitation processes. According to Cotton & Wilkinson (1988), trivalent thallium is extensively hydrolysed to form the colloidal oxide over the pH range of natural water. Depending upon the relative kinetics of reduction and hydrolysis, precipitation of thallium(III) hydroxide may be an effective mechanism for removing thallium from solution. When thallium(III) (precipitated as the oxide or hydroxide) settles into organic-rich anaerobic sediment, it will be reduced to the monovalent form, which can in turn be fixed in the sediment by reaction with sulfide to form insoluble Tl_2S (US EPA, 1978). Thallium is thus relatively depleted in seawater where thallium(III) predominates and can be enriched in sediments where organic matter accumulates under undisturbed, anaerobic conditions (Smith & Carson, 1977).

The partitioning of thallium among the water, sediment and biotic compartments of aquatic systems has rarely been investigated. In one study, however, in which thallium (100 μg/litre as thallium(I) nitrate) was added to a 7-litre glass aquarium containing washed sea sand, goldfish and submergent aquatic angiosperms, thallium was distributed among all of the compartments. Concentrations in water decreased gradually, while those in the fish and vegetation increased, throughout the 9-day duration of the experiment, indicating that thallium was being exchanged among these media (Wallwork-Barber et al., 1985). Concentrations in the sand increased rapidly to a relatively low value (0.05 mg thallium/kg), and remained relatively stable thereafter, suggesting that there was little exchange between the sediment and the other compartments. The limited accumulation of thallium in the sediment was attributed in part to the short duration of the study and to the absence of organic matter and clay in the sand.

4.1.2 Soil-vegetation transfer

4.1.2.1 Factors affecting soil-vegetation transfer

In general, the solubility of thallium compounds governs the availability of the metal to vegetation (discussed in detail by Cataldo & Wildung, 1978). Crössmann (1984) mentioned that so far no method had been developed to quantify the amount of thallium in soil that is easily available for plants. However, Schoer & Nagel (1980) emphasized the good correlation between soil-vegetation transfer and the concentration determined following ammonium acetate extraction from soil. Other authors favour an EDTA/ammonium acetate extraction (Scholl & Metzger, 1982).

Transfer is influenced by various factors, e.g., pH (section 5.1.3.2) and the type of the contaminated soil. Green rape, bush beans and rye grass were found to take up less thallium from weakly acidic soil (pH 6.2) than from more acidic soil (pH 5.6), and thallium supplied by cement factory dust was more available to plants than thallium in soil (Makridis & Amberger, 1989a). Rape plants grown on two samples of soil from a contaminated area, one sample (A) containing a 3-fold higher concentration of thallium than the other, showed identical concentrations of thallium, while other vegetables grown on sample A even showed a lower thallium content. It was concluded, that plant availability cannot be correlated to total soil thallium content as determined after extraction with concentrated nitric acid (Hoffmann et al., 1982). Only 4.4% (± 2.7%) of the thallium content of soil from a lead-zinc mining waste material area was available to vegetation, compared to 17.5% (± 10.7%) in soil from a cement plant area (Schoer & Nagel, 1980). In a similar study with soil from a cement plant and with stream sediments from a mining district (Wiesloch, Germany), rape plants took up about 20% of soil thallium from the cement plant sample but only 1.4 to 5.1% from the stream sediments, although the latter contained 2- to 3-fold higher thallium concentrations; 8- to 80-fold higher concentrations of plant-available thallium were calculated for the soil from the cement plant (Scholl & Metzger, 1982). Comparing the uptake of thallium by rape seedlings from soil contaminated by emissions from a cement plant (mainly with thallium(I) chloride or iodide) with that from uncontaminated soil (traces of thallium(I) sulfide), a 7.5-fold higher uptake from the contaminated soil was found (Lehn & Bopp, 1987).

At lower thallium concentrations, some plant species took up a higher percentage of the available thallium than at higher concentrations, perhaps in part because of the stronger toxic effects at higher concentrations. However, the total amount of thallium found in the plants and the thallium content of the artificial soil solutions were correlated, reaching up to 1000 mg/kg dry weight in green kale following one week's exposure to a concentration of 10 mg/litre (Schweiger & Hoffmann, 1983).

The transfer from soil to plant also depends on a number of factors relating to the plant, e.g., root system, kinetics of membrane transport, metabolism of thallium (Cataldo & Wildung, 1978), so that the total amount of thallium taken up is species-specific (section 5.1.4.2). This is shown by the bioconcentration factor (concentration of thallium in the plant (fresh or dry weight) in relation to its concentration in dry soil) found for different plants grown in soil contaminated by mining waste materials or collected from sites with naturally high concentrations (Table 6) (Schoer & Nagel, 1980; Lehn & Bopp, 1987). Calculations based on the concentrations in plant ash and dry soil show that the concentration factor is usually less than 20 (Smith & Carson, 1977). The concentrations of thallium in vegetables reported by these authors are one to two orders of magnitude higher than those found by Geilmann et al. (1960) in vegetation grown on uncontaminated soil (Schoer & Nagel, 1980) (sections 5.1.4.1 and 5.1.4.2). Trees can be a long-term reservoir of thallium. As a result of emission by cement plants, the bark and lichens of several trees contained 2-23.8 mg thallium/kg dry weight. The use of ground-up bark from these trees for mulching can lead to considerable uptake of thallium by other plants (Arndt et al., 1987).

4.1.2.2 Absorption by plants

Uptake of thallium(I) ions occurs via all parts of the plant, presumably by using the uptake mechanisms for potassium. However, uptake of fly-dust by the leaves of sunflowers is minimal (Schweiger & Hoffmann, 1983). Although the majority of the thallium-containing particles have a diameter less than 2 μm, they cannot be absorbed by transpiration through the stomata

Table 6. Bioconcentration factor for plants grown on contaminated soils

Plant	Bioconcentration factor[a] Fresh weight	Dry weight	Reference
Barley (*Hordeum vulgare*)		0.14	Lehn & Bopp (1987)
Cabbage species:			
Green kale	< 0.1		Schoer & Nagel (1980)
Brussels sprouts	< 0.1		Schoer & Nagel (1980)
Celeriac (*Apium graveolens*)	< 0.1		Schoer & Nagel (1980)
Cress (*Lepidium sativum*)		33	Lehn & Bopp (1987)
Cress (*Lepidium sativum*)	0.45-0.59		Schoer & Nagel (1980)
Horse-radish (*Armoracia)*	0.33		Schoer & Nagel (1980)
Maize (*Zea mays*)		0.05	Lehn & Bopp (1987)
Mushrooms	2.9		Schoer & Nagel (1980)
Mustard (*Sinapis alba*)		1.07	Lehn & Bopp (1987)
Parsley (*Petroselinum crispum lapathifolia*)	0.15-0.21		Schoer & Nagel (1980)
Rape (*Brassica napus*)		66	Lehn & Bopp (1987)
Rape (*Brassica napus*)	0.26-0.29		Schoer & Nagel (1980)
Spinach		594	Maier-Reiter et al. (1987)
Wheat (*Triticum aestivum*)		0.05	Lehn & Bopp (1987)

[a] Concentration of thallium in fresh or dry weight of the plant in relation to its concentration in dry soil

(Pallaghy, 1972; LIS, 1980). In numerous laboratory studies using nutrient solutions, a positive correlation between plant uptake and thallium concentration in the solution has been demonstrated (e.g., Al-Attar et al., 1988). Comparable results have been obtained from the cultivation of mycelium of higher fungi in thallium-enriched agar medium (Seeger & Gross, 1981).

According to Cataldo & Wildung (1978), absorption of thallium by plants seems to be under metabolic regulation, and potassium is a non-competitive inhibitor. Sunflowers with a deficiency of potassium and supplied with 1 or 10 mg thallium nitrate/litre possessed a 2 to 3 times higher concentration of thallium per gram dry weight than those supplied with potassium (Schweiger & Hoffmann, 1983). Metabolically controlled uptake seemed to occur only with thallium(I), supplied as the acetate, while thallium(III), supplied as the chloride, was presumably taken up by passive processes such as cation exchange (Logan et al., 1983, 1984). Since increasing concentrations of potassium decrease the uptake of thallium(I), this uptake was postulated to be mediated by the (Na^+/K^+) ATPase system. During a 3-h exposure to a solution concentration of 5 mg/litre, excised barley seed roots took up about 6009 (± 185) mg thallium(I)/kg dry weight and only 870 (± 44) mg thallium(III)/kg. Thallium(III) ions were easily desorbed, presumably because of a large extracellular component, whereas the thallium(I) ions were unavailable for exchange. The different uptake mechanisms are also reflected in the sensitivity of thallium(I), but not of thallium(III), towards temperature and metabolic inhibitors. Using whole plants (maize), the differences in uptake could not be confirmed, but the authors suggested that, prior to the uptake, thallium(III) may be reduced in the soil to thallium(I) (Logan et al., 1984).

4.1.2.3 *Distribution in plants*

Thallium distribution at the cellular level has been investigated with rape grown both on uncontaminated soil and on soil spiked with non-toxic amounts of thallium (Günther & Umland, 1989). At each test concentration about 70% of the thallium was concentrated in the cytosol (comparable to human data in section 6.6). In the exposed plants nearly all the thallium was in the form of free thallium(I) ions; no thallium(III) ions or dimethylthallium compounds were detected. However, in all the rape grown on uncontaminated soil, the cytosolic thallium was bound, probably to a peptide. This native thallium-complexing agent lacked sulfur-containing amino acids and could not be induced in rape by the application of thallium (Günther & Umland, 1989).

In addition to its varied distribution at the subcellular level, thallium distribution in green plants depends on the developmental stage and the part of the plant. Only in mushrooms was no specific distribution pattern found to exist (Seeger & Gross, 1981). Rape seedlings grown on soil contaminated by a cement plant (1 to 3 mg thallium/kg dry soil) contained 3 to 5 times higher concentrations of thallium than full-grown plants. The concentrations in different parts of full-grown rape (leaf, 47 mg/kg dry weight; shoot, 5.5 mg/kg; seed, 2.1 mg/kg) (Lehn & Bopp, 1987) indicate that thallium concentrations are higher in the chlorophyll-containing regions, a fact also known from plants grown on uncontaminated soils (Weinig & Zink, 1967). In rape grown on artificially contaminated soil (1 mg thallium nitrate/plant), yellowing leaves showed higher concentrations (up to 200 mg/kg dry weight) than green leaves, while the seeds contained only about 1 to 2% of the concentration found in the yellow leaves. However, in rape grown in the field near a cement plant, the leaves contained up to 85 mg thallium/kg dry weight and the seeds about 20 mg/kg (Arndt et al., 1987).

Experimentally, thallium concentrations of 0.0001 to 2.5 mg/litre substrate increased the concentration in the shoots of the grass *Lolium perenne* from < 0.075 mg/kg dry weight to 144.05 mg/kg and in the roots from 0.42 to 576 mg/kg (Al-Attar et al., 1988).

The distribution of thallium also varies in different vegetables. For instance, in gardens around Lengerich, leaves of kohlrabi contained a 350-fold higher concentration than the tubes, while in other vegetables the differences in concentrations between leaves and other parts ranged from 3 to 45 times (see Table 13) (Hoffmann et al., 1982). In studies with bush beans and green rape, differences in thallium accumulation in the plants were evident (Makridis & Amberger, 1989b): after incubation in a liquid culture medium (1 mg thallium(III) trichloride/litre) for 10 days, roots and shoots of beans contained 742 and 62 mg/kg and those of rape 57 and 244 mg/kg, respectively. At higher concentrations the difference between roots and leaves disappeared in both species, the concentration in the roots of rape increasing more strongly than in the shoots, which, in part, was an effect of reduced growth. Kaplan et al. (1990), using thallium(I) sulfate (0.55 and 1 mg/litre), observed at least 4-fold higher concentrations of thallium in the roots of soya beans than in the pods or the lower or higher leaves.

These data indicate that plants which are more resistant to thallium do not have a reduced uptake, but a reduced transport of thallium to the leaves (section 9.3.1.6).

4.2 Biotransformation

Laboratory experiments indicate that organothallium derivates may originate from the biomethylation processes of anaerobic bacteria in lake sediments (Manzo & Sabbioni, 1988). However, according to Craig (1980), there is no firm evidence for environmental methylation. The methylation of thallium and other heavy metals is a vitamin B_{12}-(cobalamin-)dependent reaction (Hill et al., 1970; Agnes et al., 1971). Due to its reduction potential, thallium(III) is methylated by methylcobalamin (Ridley et al., 1977). Transfer of the methyl group to thallium(III) seems to occur by electrophilic attack of the Co-C bond (Wood et al., 1978; Wood, 1984, 1987).

Monovalent thallium seems to be simultaneously oxidized and methylated by specific anaerobic microorganisms to methylthallium(III) moieties which are stabilized by complexation (Huber et al., 1978).

Oxidation of thallium(I) ions to thallium(III) oxide in yeast mitochondria (Lindegren, 1971; Lindegren & Lindegren, 1973b) confirms an *in vivo* oxidation, but specific culture conditions are necessary to obtain this detoxification phenomenon in which thallium oxide is deposited between cell wall and plasma membrane.

4.3 Interaction with other physical, chemical, or biological factors

In the atmosphere, chemical reactions involving thallium are not very likely to occur (Schoer, 1984).

5. ENVIRONMENTAL LEVELS AND HUMAN EXPOSURE

5.1 Environmental levels

Because of the limited industrial uses of thallium, emission on a global scale resulting from the production and use of thallium compounds is unlikely. However, thallium is present in relatively large amounts in the raw materials used in various industrial processes (e.g., smelting of sulfide ores, power generation using coal, brick and cement manufacturing) (Table 5), which when released can significantly increase environmental exposure to thallium on both a local and regional scale.

5.1.1 Air

Bowen (1979) reported mean values of 0.06 ng particulate thallium/m^3 air for Europe and 0.22 ng/m^3 for North America, and Arnold (1986) a range of 0.1 to 30 ng/m^3. The air of six large American cities contained < 0.04 to 0.1 ng thallium/m^3 (Ohnesorge, 1985). In a detailed study at Chadron, Nebraska, USA, Struempler (1975) found yearly mean values for thallium of 0.22 ± 0.08 (range 0.07 to 0.48 ng/m^3) and of 0.15 ± 0.04 ng/m^3 during the summers of 1973 and 1974, respectively.

In industrial and urban areas of Genoa, Italy, the geometric mean concentrations of thallium have been found to be 15 and 14 ng/m^3 air, respectively, with maximal values of 58 ng/m^3, but often values were below 1 ng/m^3 (Valerio et al., 1988, 1989). In London, levels of 0.07 to 6 mg thallium/kg dust were measured (Bowen, 1979).

Air emission by thallium is mainly caused by mineral smelters, power-generating plants and cement plants (ATSDR, 1992). Thallium compounds are volatile at high temperatures and are not efficiently retained by most emission control facilities. In consequence, large amounts of thallium are released into the atmosphere if the raw material (coal or ores) is not selected for a low thallium content.

In fly-ash from a power plant, only 0.08% of particles were < 5 μm in diameter with a thallium concentration of 45 mg/kg ash (Natusch et al., 1974).

5.1.2 Water

In the majority of reports, the authors did not specify whether they determined dissolved and/or particulate thallium in samples of water. This information was included if possible, according to the methodology used in the study.

5.1.2.1 Areas not contaminated by thallium

Seawater contains < 0.01 to 0.02 μg/litre, and river water < 0.01 to 1 μg/litre (Mason, 1966; Smith & Carson, 1977; Bowen, 1979; Kemper & Bertram, 1984; Wachs, 1988).

In volcanic springs, low concentrations have been found (0.25 μg/litre) (Arnold, 1986) and in three samples of hydrothermal water it was below the detection limit of 0.6 μg/litre (Korkisch & Steffan, 1979). Although, in these and the wastewater investigation, dissolved and particulate thallium were not determined separately, Henshaw et al. (1989) found concentrations of up to 0.41 μg thallium/litre in filtered water from freshwater lakes. In three wastewater treatment facilities in Massachusetts, USA, that had no major industrial waste inputs, the thallium concentration in the influent was below the detection limit of 5 μg/litre (Aulenbach et al., 1987).

However, wastewater from oilfields (oil-well brines) in the USA contained 12.9 to 672 μg thallium/litre, 5 out of 13 samples containing > 400 μg thallium/litre (Korkisch & Steffan, 1979).

5.1.2.2 Areas contaminated by thallium from industrial sources

Data on thallium emission in water are available for areas with oilfields, mineral industry and cement plants (Table 7).

Increased concentrations of thallium in well water and in water from an irrigation canal in China resulted from old waste materials from the mining of mercury and coal (Zhou & Liu, 1985) (section 5.1.3.2). Effluents from tailing ponds of base-metal mining operations in New Brunswick, Canada contained 27 and 1620 μg dissolved thallium/litre and up to 88 μg/litre was found in connecting rivers (Zitko et al., 1975; Zitko, 1975b).

Raw wastewater from a pyrite ore mine at Lennestadt, Germany (which was the source of the pyrite roasting residues used by the cement plant in Lengerich, Germany) showed a

Table 7. Concentrations of thallium in water from contaminated areas[a]

Locality	Source	Concentration of thallium (µg/litre)	Reference
Underground water, Düsseldorf, Germany	zinc smelter	13-820	BGA (1979)
Wastewater, different locations, USA	smelters[b]	< 0.1-2400	Smith & Carson (1977)
Tailing ponds, New Brunswick, Canada	mining	27; 1620	Zitko et al. (1975)
Rivers, different locations, USA	mining	21-30	US EPA (1980)
Rivers, New Brunswick, Canada	mining	1-88	Zitko et al. (1975)
River, Lennestadt, Germany	mining	< 1-1	LIS (1980)
Wastewater, Lennestadt, Germany	mining	2-160	LIS (1980)
Well, China	mining	17-40	Zhou & Liu (1985)
Irrigation canal, China	mining	6-96	Zhou & Liu (1985)
Wells, Lengerich Germany	cement plant	< 1	LIS (1980)
Surface water, Lengerich, Germany	cement plant	< 1-1	LIS (1980)
Water, Lengerich, Germany	cement plant (distance 1 km)	7.3	Mathys (1981)
Wastewater, Lengerich, Germany	cement plant	< 1-37	LIS (1980)
Wells, Erwitte etc., Germany	cement plant	< 5-< 50	LIS (1980)
Surface water, Erwitte etc., Germany	cement plant	< 50-130	LIS (1980)
Wastewater, Erwitte etc., Germany	cement plant (flue dust)	800	LIS (1980)

Table 7 (contd).

Locality	Source	Concentration of thallium (μg/litre)	Reference
Wastewater, USA	oil drilling (natural brines)	12.9-672	Korkisch & Steffan (1979)
Wastewater, Canada	iron and steel plants	mean = 60	MISA (1991a)
	pulp and paper mills	230; mean = 52	MISA (1991b)
	petroleum refineries	310; mean = 19	MISA (1989)

[a] Further literature summarized by Schoer (1984)
[b] Detailed data of sulfide mineral processors in Smith & Carson (1977)

thallium concentration of 160 μg/litre (LIS, 1980). Treatment in sedimentation ponds and with lime and chlorine, reduced the concentration to 2-35 μg/litre. In a stream, used as main drainage channel, the thallium concentration rose from < 1 μg/litre (detection limit) to 1 μg/litre after the inlet. This level is similar to that found in the River Rhine, Germany (0.5 and 2.5 μg/litre) (LIS, 1980).

Groundwater directly below a depot for pyrite roasting residues in Duisburg, Germany contained 17 μg thallium/litre, and, at a distance of some 100 m, up to 6 μg/litre (LIS, 1980).

In the vicinity of the cement plant in Lengerich, Germany (see section 4.2.1.4) thallium levels were monitored in wells, rivers and wastewater (LIS, 1980). In rivers, levels of 7.3 μg/litre 1 km from the plant decreased to 1.8 μg/litre at a distance of 5 km and to 1.0 μg/litre at a distance of 10 km (Mathys, 1981). In all private wells and water works, thallium concentrations were below the detection limit of 1 μg/litre. At the purification plant of the cement plant in Lengerich, the water from the inlet showed a concentration of 17 μg/litre and that from the outlet of the rainwater collection pond 37 μg/litre. A rainwater pond used for watering cattle contained 3 μg thallium/litre.

Around other cement plants, thallium concentrations in private wells, water works and surface water were below the detection limit (< 5 and < 50 μg/litre). Pond water from the vicinity contained 130 μg/litre. In drip-water from the storage of flue dust, a concentration of 800 μg/litre was determined (LIS, 1980).

The elimination of thallium from wastewater varies. Only 28% of the thallium could be removed by conventional wastewater treatment (liming) (Zitko et al., 1975) whereas 80 to 98% was removed in Lennestadt (LIS, 1980). It has been suggested that Prussian Blue could be used to eliminate thallium from wastewater (Rauws & Canton, 1976). Wool cannot be used as a filter to remove thallium from contaminated water, since, in contrast to other metallic ions, only minor amounts of thallium are adsorbed (Masri, 1976).

5.1.3 Rocks, soil and sediment

5.1.3.1 Areas not contaminated by thallium

Mean concentrations in the earth's crust range from 0.1 to 1.7 mg/kg. Higher values (up to 3 mg/kg) have been determined for granite and shale; intermediate values for basalt, limestone, sandstone and most coals, and lowest values for dunite (Table 8) (Mason, 1966; Bowen, 1966, 1979; Brumsack, 1977; Smith & Carson, 1977; Kemper & Bertram, 1984, 1991; Schoer, 1984; Arnold, 1986). Much higher concentrations can occur in organic-rich shales such as the Pierre Shale in the USA (25 mg/kg) and in coals of the Jurassic period in Tadzhikistan (100 to 1000 mg/kg) (Smith & Carson, 1977).

Total thallium concentrations in soil typically range from 0.1 to about 1.0 mg/kg (Geilmann et al., 1960; Bowen, 1966, 1979; Chattopadhyay & Jervis, 1974; Brumsack, 1977; Smith & Carson, 1977; Schoer, 1984; OMEE, 1994), but in China are around 0.011 mg/kg in garden soil (range 0 to 0.02 mg/kg) (Zhou & Liu, 1985). Higher concentrations (up to 5 mg total thallium/kg) have been reported, however, in Poland (Staszyc et al., 1986), in soil on shale (Hoffmann et al., 1982) and near some metallic ore deposits (Smith & Carson, 1977).

Marine sediments have been found to contain 0.95 mg thallium per kg (Bowen, 1979) or, according to McLaren et al. (1987), using isotope dilution inductively coupled plasma mass spectrometry, 0.6 to 0.7 mg/kg. Data summarized by Smith & Carson (1977) show

Table 8. Concentrations of thallium in uncontaminated geological samples[a]

Source	Concentration of thallium (mg/kg)		Reference
	Mean	Range	
Basalt		< 0.2-0.7	Smith & Carson (1977)
Basalt	0.08		Bowen (1979)
Basalt		0.02-0.06	Arnold (1986)
Clay		0.3	Smith & Carson (1977)
Clay		440-470[b]	Smith & Carson (1977)
Clay	0.9		Bowen (1979)
Coal	0.38	< 0.2-1.4	Smith & Carson (1977)
Coal	0.2	0.01-2	Bowen (1979)
Coal	0.6	0.12-1.3	Gluskoter et al. (1977)
Brown coal (18% ash)	0.027		Brumsack et al. (1984)
Hard coal (8.7% ash)	0.51		Brumsack et al. (1984)
Hard coal (13.9% ash)	0.72		Brumsack et al. (1984)
Dunite	0.0005		Bowen (1979)
Granite	3.1	0.3-6.4	Smith & Carson (1977)
Granite	1.1		Bowen (1979)
Limestone	1.7		Smith & Carson (1977)
Limestone	0.14		Bowen (1979)
Limestone		0.1-0.9	Arnold (1986)
Sandstone	0.8		Smith & Carson (1977)
Sandstone	0.36		Bowen (1979)
Sandstone		0.05-0.4	Arnold (1986)
Shale (low in organic carbon)	0.68		Brumsack et al. (1984)
Shale	1.2		Bowen (1979)
Shale	3.1		Smith & Carson (1977)
Black shale (rich in organic carbon)	25		Smith & Carson (1977)

[a] Selected data; detailed data summarized in Smith & Carson (1977); Bowen (1966, 1979); Schoer (1984)

[b] Very fine inclusions of plant matter

a range of 0.14 to 1.13 mg/kg and in manganese nodules up to 614 mg/kg.

Uncontaminated sediments from lakes and small streams in various parts of Canada typically contain about 0.35 mg thallium per kg (range 0.02 to 3.2 mg/kg) (G. Bonham-Carter, Geological Survey of Canada, Applied Geochemistry Subdivision, personal communication to the IPCS), with lowest values occurring in areas with underlying basaltic rock. Thallium levels in the sediment of small streams in an uncontaminated area of Münsterland, Germany contained from 0.03 to 0.1 mg/kg (LIS, 1980). In another investigation of small rivers in the same area and in Sauerland, Germany, concentrations of 0.01 to 0.07 mg/kg dry weight were determined (Mathys, 1981). This is also in the range of the data summarized by Smith & Carson (1977).

5.1.3.2 *Areas contaminated by thallium from industrial sources*

Cases of contamination of sediment and soil by thallium are mainly caused by mineral mining and smelters and by dust fall-out from emissions of power-generating plants, brickworks and cement plants (ATSDR, 1992) (Table 9).

Emissions from the cement plant in Lengerich, Germany caused a remarkable increase in thallium concentrations in sediments of rivers and brooks (Mathys, 1981). Sediment levels of 18 mg thallium/kg dry weight found in a brook 1 km from the plant decreased to 8.7 mg/kg within 4 km and then to 7.5 mg/kg in the River Glane into which the brook flowed. Sediments of the following River Ems contained 5.0, 2.7 and 0.8 mg/kg at distances of 30, 70 and 100 km, respectively. In comparison, river sediments from industrialized areas contained 0.05 to 1.8 mg/kg dry weight. Very high thallium levels were detected in sediments from areas with zinc mining or iron ore industry, e.g., 40.0 mg/kg in the River Lenne. After transport to the River Ruhr, sediment thallium levels were 3 mg/kg dry weight (Mathys, 1981).

Large amounts of contaminated waste materials from the mining of mercuric ore and coal containing 25 to 106 mg thallium per kg resulted in chronic thallium poisoning in China. As a result of dispersal of the waste materials, the garden soil of poisoned owners showed levels from 28 to 61 mg/kg (mean: 43 mg/kg), whereas the soil levels in gardens of unaffected families contained 6 to 11 mg/kg (mean, 8 mg/kg); this was still much higher than in other villages (mean, 0.011 mg/kg). In the affected village, the concentration of soluble thallium salts decreased with increasing pH, with 8.0, 2.0 and < 0.15 mg/kg at soil pH values of 1-2, 3-4

Table 9. Concentrations of thallium in soil from the vicinity of factories in Germany

Locality	Source	Distance (m)	Concentration of thallium (mg/kg)	Reference
Göttingen	lead-zinc smelter	-[a]	1.07	Brumsack (1977)
Duisburg	copper smelter	500-1400	< 0.2-2.1	LIS (1980)
Leimen, Wiesloch	mining and cement plant	-[a]	5.5-21	Hoffmann et al. (1982)
Lengerich	cement plant	500-5000	< 0.1-6.9	LIS (1980)
Erwitte	cement plant	350-1800	0.1-10.5	LIS (1980)
Schelklingen[b]	cement plant	< 3200	0.1-0.5	Arndt et al. (1987)
Mergelstetten[b]	cement plant	< 800	0.1-0.5	Arndt et al. (1987)
Duisburg	sulfuric acid plant	0-1000	< 0.2-10.5	LIS (1980)
Duisburg	sulfuric acid plant	350-1100	< 0.2-2.3	LIS (1980)
Göttingen	brickwork	-[a]	0.6	Brumsack (1977)

[a] No data given;

[b] The investigation was performed 6 years after a ban on the use of iron pyrite residues with high thallium contaminations. Owing to the method used (acid digestion with concentrated nitric acid for 2 h at 90-95 °C and addition of 10% sulfuric acid), the concentrations measured correspond to the extractable soluble emitted thallium and not the total thallium in the soil.

and 6-7, respectively. The lower soil pH in the dry season (3.5-4.5 compared to pH 6-7 in the rainy season) correlated with an increase in the number of intoxications during these months, presumably due to an increase in the thallium concentration in cabbage. After the experimental addition of lime to contaminated soil, the thallium concentration in cabbage was reduced (Zhou & Liu, 1985).

Sabbioni et al. (1984b) calculated the emissions from a hypothetical coal-fired power-generating plant for a period of 40 years. They deduced an air-borne deposition of thallium around the power-generating plants of 0.005 mg/kg, and the factor of increase over the background level was estimated to be 0.001.

Around four small brickworks, samples of soil were digested by strong acids and analysed for total concentration of thallium (Brumsack, 1977). The proportion that was bioavailable is unknown. Compared to a soil background level of 0.2 mg/kg, the contaminated samples of soil showed a maximum accumulation factor of 3, while for samples taken directly around the factory the factor was about 5. Clear effects were found when the weather side of a hill was just opposite the smoke stack. (Interestingly, shale from uncontaminated areas showed a similarly high content (0.99 mg/kg)).

Thallium emission by the cement plant in Lengerich, Germany caused an increase of thallium concentrations in the soil over an area of 1 to 2 km radius from the plant, with a maximal level of 6.9 mg/kg dry soil (LIS, 1980). Up to 4 mg/kg soil was determined in agricultural soil and up to 6 mg/kg in the soil of house gardens (Crössmann, 1984). Samples of soil taken at different depths always showed highest thallium contaminations in the upper layers, decreasing with increasing depth (LIS, 1980). Only small amounts of the thallium in the upper layers were washed out (section 4.1) (Scholl & Metzger, 1982). The soil around the two plants that had produced the residues from pyrite roasting was also highly contaminated, with maximal levels of up to 10.5 and 2.3 mg/kg, respectively (LIS, 1980).

Soil from the vicinity of two other cement plants in Germany contained only slightly elevated concentrations of thallium, up to 0.5 mg/kg soil, in the upper layers (see section 3.2.3.3 and 5.1.4.2) (Table 9) (Arndt et al., 1987).

5.1.4 Plants and animals

Thallium occurs in low amounts in almost all living organisms, including humans (Mason, 1966) (Tables 10 and 14). It seems to be a non-essential cation in animals and plants (Yopp et al., 1974). Some species accumulate this element.

5.1.4.1 *Plants*

a) *Areas not contaminated by thallium*

Usually thallium concentrations in plants are much less than 0.1 mg/kg dry weight (Geilmann et al., 1960) or 1 mg/kg ash (Dvornikov et al., 1973, 1976), and levels exceeding 2 mg/kg ash are unusual (Smith & Carson, 1977) (Table 10). However, such high thallium concentrations have been found in plants from areas with a naturally very high thallium concentration, e.g., the Alsar in Macedonia, Yugoslavia (Zyka, 1972) (Table 11). Data from this area are considered in sections 5.1.4.2 and 9.3.1.

No thallium could be detected in cabbage or grain from areas of China not contaminated by thallium (Zhou & Liu, 1985). Plants used for teas (e.g., *Anisi, Betulae, Hibisci* and *Menthae*) contained very low concentrations of thallium (< 0.01 mg/kg), whereas higher concentrations of other heavy metals and pesticides often occurred (Ali & Blume, 1983). The majority (85%) of 421 investigated species of wild mushrooms, which often accumulate heavy metals, contained concentrations below the detection limit of < 0.25 mg/kg dry weight (range < 0.25 to 5.5 mg/kg). A transfer factor of < 0.1 (concentration of thallium in the mushroom (fresh or dry weight) in relation to its concentration in dry soil) indicates that no accumulation took place (Seeger & Gross, 1981). However, in other plants and using soil from a contaminated area, a much higher transfer factor of 2.9 was determined (Schoer & Nagel, 1980). The thallophilic Brassicaceae can contain higher amounts of thallium (1.5 mg/kg fresh weight) than other plants, which usually contain 0.007 (detection limit) to 0.1 mg/kg wet weight (Crössmann, 1984).

Wild plants normally contain only traces of thallium, whereas the levels in garden plants can be increased by repeated use of sewage sludge or potash fertilizers, which can contain 100 to 210 μg/kg sludge or 15 to 310 μg/kg fertilizer (Geilmann et al., 1960; Heinrichs, 1982). Also phosphate and copper fertilizers may contain up to 400 μg thallium/kg (Boysen, 1992).

Thallium concentrations of up to 17 g/kg ash have been found in plants from Alsar in Macedonia, Yugoslavia (Table 11), an area with very high geogenic thallium levels (Zyka, 1972).

Table 10. Concentrations of thallium in plants from uncontaminated areas[a]

Source[b]	Concentration of thallium (μg/kg dry weight)	Concentration of thallium (mg/kg ash)	Reference
Achillea millefolium		0.01-0.04	Dvornikov et al. (1973)
Achillea setacea		0.04-0.9	Dvornikov et al. (1976)
Alpine fir (L)		2-100	Shacklette et al. (1978)
Alpine fir (S)		2-70	Shacklette et al. (1978)
Anthemis tinctoria		< 0.1-0.5	Dvornikov et al. (1976)
Artemisia absinthum		0.03	Dvornikov et al. (1973)
		0.02-0.6	Dvornikov et al. (1976)
Artemisia campestris		0.057	Dvornikov et al. (1973)
		0.04-0.8	Dvornikov et al. (1976)
Asperula humifusa		0.1-1.0	Dvornikov et al. (1976)
Clover	8-10		Geilmann et al. (1960)
Echium vulgare		0.1-0.3	Dvornikov et al. (1976)
Endive	80		Geilmann et al. (1960)
Engelmann's spruce (L)		2-10	Shacklette et al. (1978)
Engelmann's spruce (S)		15	Shacklette et al. (1978)
Euphorbia virgata		0.022-0.027	Dvornikov et al. (1973)
		0.03-0.3	Dvornikov et al. (1976)
Festuca sulcata		0.2	Dvornikov et al. (1973)
		0.2-0.6	Dvornikov et al. (1976)
Green cabbage	125		Geilmann et al. (1960)
Hay	20-25		Geilmann et al. (1960)
Head-lettuce	21		Geilmann et al. (1960)
Herbaceous vegetables	30-300		Bowen (1979)
Kale	150		Bowen (1979)
Leek	75		Geilmann et al. (1960)
Limber pine (L)		2-5	Shacklette et al. (1978)
Limber pine (S)		3-5	Shacklette et al. (1978)
Lodgepole pine (L)		2-5	Shacklette et al. (1978)
Lodgepole pine (S)		3-7	Shacklette et al. (1978)
Mushrooms	< 0.25-5.5		Seeger & Gross (1981)
Myrtle blueberry (L, S)		2-7	Shacklette et al. (1978)
Ponderosa pine (S)		15	Shacklette et al. (1978)
Potato (L, S)	25-30		Geilmann et al. (1960)
(T)	5		Geilmann et al. (1960)
Rape (L)	25-30		Geilmann et al. (1960)
Red cabbage	40		Geilmann et al. (1960)

Table 10 (contd).

Source[b]	Concentration of thallium (μg/kg dry weight)	(mg/kg ash)	Reference
Salvia nemorosa		0.04	Dvornikov et al. (1973)
		0.04-0.8	Dvornikov et al. (1976)
Stinging nettle (L)	28.8		Weinig & Zink (1967)
Tanacetum vulgare		0.06-0.2	Dvornikov et al. (1976)
Tobacco (L)	24-100		Geilmann et al. (1960)
Verbascum ovalifolium		0.01-0.7	Dvornikov et al. (1976)
Woody gymnosperms	50		Bowen (1979)

[a] Further data summarized by Dvornikov et al. (1973, 1976), Gough et al. (1979) and Smith & Carson (1977)

[b] L = leaves, needles; S = stems; T = tubers

Table 11. Concentrations of thallium in plants from the Alsar region in Yugoslavia possessing a high natural concentration of thallium in the soil

Source	Concentration of thallium (mg/kg ash weight)
Campanula sp. (L, S, F)	5990
Centaurea sp. (P)	75
Centaurea sp. (L, S)	105
Dianthus sp. (F)	5200
Echinops sp. (L)	15
Eryngium sp. (L)	3
Eryngium sp. (F)	10
Galium sp. (F)	17 000
Lavatera sp. (L, S)	125
Lavatera sp. (F)	45
Linaria triphylla	3000
Linaria triphylla	3800

According to Zyka (1972)

F = flowers; L = leaves, needles; P = pods and seeds; S = stems

b) Areas contaminated by thallium from industrial sources

Soils contaminated through mineral smelters, power-generating plants, brickworks or cement plants can greatly increase the concentrations of thallium found in food of plant origin (Tables 12 and 13), which are the major route of entry of thallium into the food chain. Data on bioconcentration factors are listed in Table 6. In the contaminated area of Lengerich, Germany, consumption of home-grown food was correlated with high levels of thallium in urine and hair, and possibly with thallium-related health disorders among local people (Brockhaus et al., 1981b; Dolgner et al., 1983). The importance of these findings is underlined by the similarly elevated levels of thallium found in the urine of family members consuming the home-grown vegetables (Ewers, 1988).

Waste materials from the mining of mercuric ore and coal in China (section 5.1.3.2) increased the concentration of thallium in cabbage and grain. Cabbage from gardens of affected families contained 42 mg/kg fresh weight (range 39 to 49), whereas cabbage eaten by healthy families contained 5.6 mg/kg (range 3 to 11). Other vegetables from the gardens of affected families usually contained less than 10 mg/kg (Zhou & Liu, 1985).

An accumulation in vegetables of the genus *Brassica* was also observed in Lengerich. In the area with the highest contamination, the majority of plants and fruits contained < 0.1 to 0.4 mg/kg fresh weight. Higher thallium levels were sometimes found in strawberries, potatoes, beans, tomatoes, carrots and leeks, while in parsley, celery, red currants, and all Brassicaceae high levels were usual (LIS, 1980). Within this genus uptake of thallium varied: the bioconcentration factor of white and red cabbage was relatively low; it was 5- to 10-fold higher in the stems of kohlrabi. Savoy cabbage and green kale were found to contain the highest thallium concentrations, exceeding those of the soil (Crössmann, 1984). The maximal value of 45.2 mg/kg fresh weight was found in green kale (LIS, 1980). Most forage plants, e.g., turnips, hay, grass and fodder corn, contained < 5 mg/kg dry weight, but 46% of rape plants contained > 100 mg/kg (up to 1095 mg/kg dry weight) and 22% of the maize 10 to 50 mg/kg (LIS, 1980). Vegetables from Lengerich, grown in soil with 4.5 mg thallium/kg dry weight, could be classified according to their mean thallium concentration (mg/kg fresh weight) into five groups. These were I: green cabbage (22.6 mg/kg) and savoy cabbage (8.5 mg/kg); II: turnip, broccoli, kohlrabi and white cabbage (3.1 mg/kg); III: stock beet and other vegetables (1.4 mg/kg); IV: red beet, rhubarb and

Table 12. Concentrations in plants from thallium-contaminated areas

Organism	Concentration of thallium (mg/kg)		Source of emission	Locality[a]	Reference
	Dry weight	Fresh weight			
Algae	9.5-43.4		mining	New Brunswick	Zitko et al. (1975)
Algae		0.665	cement plant	Lengerich	LIS (1980)
Berula	100.3		cement plant	Lengerich	Mathys (1981)
Berula		0.585; 0.654	cement plant	Lengerich	LIS (1980)
Caltha	187.3		cement plant	Lengerich	Mathys (1981)
Elodea	87.4		cement plant	Lengerich	Mathys (1981)
Elodea		0.29; 6.5	cement plant	Lengerich	LIS (1980)
Grass	52.0		sulfuric acid plant	Duisburg	LIS (1980)
Moss	125; 162		mining	New Brunswick	Zitko et al. (1975)
Rape		29.2	cement plant	Lengerich	LIS (1980)
Rape		23.7	cement plant	Lengerich	Kemper & Bertram (1984)
Rape	1095		cement plant	Lengerich	LIS (1980)
Rape	679.6		cement plant	Mergelstetten	Arndt et al. (1987)
Rape	19.1		cement plant	Schelklingen	Arndt et al. (1987)
Sparganium		0.265	cement plant	Lengerich	LIS (1980)

[a] All the localities are in Germany, except for New Brunswick (Canada)

Table 13. Concentrations in vegetables and fruits from thallium-contaminated areas

Plant	Part	Concentration of thallium (mg/kg)[a]		Source of emission	Reference
		Dry weight	Fresh weight		
Apple	fruit		0.2	cement plant	LIS (1980)
Bean	fruit		0.7	cement plant	LIS (1980)
Blackberry	fruit		0.5	cement plant	LIS (1980)
Black-currant	fruit		0.527	cement plant	Kemper & Bertram (1984)
Brussels sprout	leaf		0.5	cement plant	LIS (1980)
Carrot	root		1.0	cement plant	LIS (1980)
Carrot	leaf root	0.30 0.10		mining and cement plant	Hoffmann et al. (1982)
Celeriac	stem		0.8	cement plant	LIS (1980)
Cucumber	leaf fruit	0.70 0.10		mining and cement plant	Hoffmann et al. (1982)
Green cabbage	leaf		14.9	cement plant	Kemper & Bertram (1984)
Green cabbage	leaf		45.2	cement plant	LIS (1980)

Table 13 (contd).

Green cabbage	leaf		22.6[b]	cement plant	Scholl & Metzger (1982)
Kohlrabi	leaf	35.00		mining and cement plant	Hoffmann et al. (1982)
	stem	0.10			
Kohlrabi	stem		3.1[b]	cement plant	Scholl & Metzger (1982)
Kohlrabi	stem		4.9	cement plant	LIS (1980)
Onion	stalk	0.10		mining and cement plant	Hoffmann et al. (1982)
	bulb	0.01			
Onion	stalk		0.4	cement plant	LIS (1980)
Parsley	leaf		1.2	cement plant	LIS (1980)
Pear	fruit		0.5	cement plant	LIS (1980)
Potato	tuber		0.8	cement plant	LIS (1980)
Radish	leaf	5.90		mining and cement plant	Hoffmann et al. (1982)
	root	0.40			
Red beet	leaf	2.40		mining and cement plant	Hoffmann et al. (1982)
	root	0.60			

Table 13 (contd).

Plant	Part	Concentration of thallium (mg/kg)[a]		Source of emission	Reference
		Dry weight	Fresh weight		
Red beet	root		0.7	cement plant	LIS (1980)
Red-currant	fruit		1.1	cement plant	LIS (1980)
Savoy cabbage	leaf		8.5[b]	cement plant	Scholl & Metzger (1982)
Strawberry	fruit		0.9	cement plant	LIS (1980)
Tomato	fruit		0.6	cement plant	LIS (1980)
White cabbage	leaf		3.1[b]	cement plant	Scholl & Metzger (1982)
Zucchini	leaf	0.90		mining and cement plant	Hoffmann et al. (1982)
	stem	0.02			

[a] Individual value, unless otherwise stated
[b] Mean value

spinach (0.7 mg/kg); V: the majority of fruits and vegetables (0.5 mg/kg), e.g., red cabbage, Brussels sprouts, onion, salad, carrot, bean, tomato, cucumber and potato (Scholl & Metzger, 1982).

The accumulating capacity of rape also became evident in an investigation at cement factories in Schelklingen and Mergelstetten, Germany, 6 years after the use of the same iron pyrite residues that had been used in Lengerich was banned. The soil contained only slightly elevated concentrations of thallium (section 5.1.3.2), and in the majority of the plants, four of them Brassicaceae, no thallium was detectable (Arndt et al., 1987). However, rape contained increased levels of 2.4 to 679.6 mg thallium/kg dry weight at Mergelstetten and 1.8 to 19.1 mg/kg dry weight at Schelklingen (Table 12). The highest levels detected in single rape plants were found within an area extending 150-400 m downwind from the cement plant. The majority of the rape grown in that area contained more than 5 mg thallium/kg dry weight and could not be used as animal feed.

5.1.4.2 *Animals*

a) *Areas not contaminated by thallium*

Investigations of three species of freshwater fish, the omnivorous white sucker (*Catostomus commersoni*) and the more carnivorous yellow perch (*Perca flavescens*) and brook trout (*Salvelinus fontinalis*), show them to have similar average concentrations of thallium in their axial muscle (< 0.07 to 3.0 mg/kg dry weight), which were independent of water pH (Heit, 1985) (Table 14). Extensive studies of different marine shellfish and fish revealed average concentrations of 0.14 mg/kg; only in three species (occasionally *Clupanodon punctatus* and *Trachurus japonicus* and often *Penaeus japonicus*) were concentrations above 1 mg/kg found (Hamaguchi, 1960).

In marine invertebrates concentrations were even lower, but, owing to the low concentration in the seawater, the concentration factor (concentration in the organism divided by the concentration in the seawater) calculated by Smith & Carson (1977) was > 700.

Thallium concentrations in marine mammals have rarely been investigated. In the blubber, liver, kidney, spleen and muscle of bowhead whales, concentrations are nearly always below 0.01 mg/kg fresh weight (Byrne et al., 1985).

Table 14. Concentrations of thallium in animals from uncontaminated areas

Source	Part	Number of measurements	Concentration of thallium		Reference
			µg/kg wet weight	mg/kg dry weight	
Invertebrates					
Colorado beetle	whole	1		18	Geilmann et al. (1960)
Different marine invertebrates				0.001-0.03	Noddack & Noddack (1939)
Echinoderms (hard parts)			110		Bowen (1979)
Molluscs (soft parts)			340		Bowen (1979)
Fish					
Various fish species			80		Bowen (1979)
Various marine shellfish and fish		139	< 2930		Hamaguchi (1960)
Brook trout (*Salvelinus fontinalis*)	muscle	5		< 3.0	Heit (1985)

Table 14 (contd).

White sucker (*Catostomus commersoni*)	muscle	28		< 2.0	Heit (1985)
Yellow perch (*Perca flavescens*)	muscle	27		< 3.0	Heit (1985)
Birds					
Duck[a]	kidney	15	0.03		Holm et al. (1987)
Duck[b]	kidney	10	0.129		Holm et al. (1987)
Duck[a]	liver	15	0.022		Holm et al. (1987)
Duck[b]	liver	10	0.207		Holm et al. (1987)
Hen	liver	2	< 50		LIS (1980)
Hen	muscle	2	< 50		LIS (1980)
Mammals					
Cattle	hair	1		20	Geilmann et al. (1960)
Cattle	hoof	1		16	Geilmann et al. (1960)
Cattle	horn	1		10	Geilmann et al. (1960)

Table 14 (contd).

Source	Part	Number of measurements	Concentration of thallium		Reference
			μg/kg wet weight	mg/kg dry weight	
Fox	intestine	25	< 2.7		Munch et al. (1974)
Fox	kidney	27	0.01-1.5		Munch et al. (1974)
Fox	liver	27	0.01-1.6		Munch et al. (1974)
Goat	hair	1		7	Geilmann et al. (1960)
Goat	hoof	1		9	Geilmann et al. (1960)
Hare	hair	1		17	Geilmann et al. (1960)
Horse	hair	1		7	Geilmann et al. (1960)
Horse	hoof	1		4	Geilmann et al. (1960)
Marten	brain	7	< 0.1-0.7		Clausen & Karlog (1974)
Marten	intestine	13	< 0.01-0.57		Clausen & Karlog (1974)
Marten	kidney	17	< 0.01-3.5		Clausen & Karlog (1974)
Marten	liver	36	< 0.01-1.4		Clausen & Karlog (1974)

Table 14 (contd).

Pig	hair	1		9	Geilmann et al. (1960)
Pig	hoof	1		11	Geilmann et al. (1960)
Pig	muscle	1		0.028	Kemper & Bertram (1984)
Pig	muscle	43	< 70		Konermann et al. (1982)
Pig	kidney	43	< 70		Konermann et al. (1982)
Pig	kidney	6	< 50		LIS (1980)
Pig	liver	43	< 70		Konermann et al. (1982)
Pig	liver	6	< 50		LIS (1980)
Rabbit	hair	1		60	Geilmann et al. (1960)
Rabbit	hair	1		< 1.5	LIS (1980)
Roe deer	liver	19	~ 30		Holm et al. (1987)
Roe deer	kidney	19	~ 30		Holm et al. (1987)
Sheep	hair	1		9	Geilmann et al. (1960)
Sheep	hoof	1		12	Geilmann et al. (1960)
Sheep	kidney	3	50-60		Hapke et al. (1980)
Sheep	liver	3	< 50		Hapke et al. (1980)
Sheep	muscle	3	50-60		Hapke et al. (1980)

[a] Cuxhaven, Germany (coast)
[b] Cuxhaven, Germany (inland)

Meat from farm animals contains very low levels of thallium (Table 14).

b) Areas contaminated by thallium from industrial sources

Different animals as well as different organs vary with respect to their accumulation capacity for thallium (Table 15).

In a case of fish poisoning, three species were found to contain 77 to 96 mg/kg muscle (Palermo et al., 1983). The liver and kidneys of fish from a pond contaminated by a cement plant contained 1.6 and 1.3 mg/kg fresh weight, respectively (LIS, 1980). In the same area, fish from other ponds and waters usually contained < 0.1 mg/kg muscle.

In farm animals intake of thallium mainly occurs through contaminated feed. In broilers and laying hens, tissue thallium concentrations were linearly correlated with feed levels for concentrations between 2 to 40 mg/kg fresh weight of feed (Ueberschär et al., 1986). The accumulation factor (concentration of thallium in the tissue in relation to its concentration in the feed) was 2 to 3 times higher for the tissues of broilers than for those of hens (Table 16). In contrast to the situation in sheep, cattle and pigs, thallium accumulates in hens to a greater degree in the muscle than in the liver. The concentration in the kidneys is about 90% lower than in the egg shell. Thallium half-life is about 2 to 4 days for the various hen tissues (Ueberschär et al., 1986).

Whereas maximal permitted concentrations of lead, mercury, arsenic and fluoride in animal fodder have been established in Germany, this has not been done for thallium (Crössmann, 1985). Thallium poisoning in cattle has been caused by silage (41 mg thallium/kg fresh weight) bought from a farm in a contaminated area (Frerking et al., 1990). Thallium mainly accumulates in the kidneys, liver and bones (section 6.2). Steers fed for at least 6 months with fodder originating from the thallium-contaminated area around Lengerich, Germany, containing about 1.25 mg/kg dry weight (daily uptake: 0.025 mg/kg body weight), contained 0.10 ± 0.02, 1.66 ± 0.55, 0.52 ± 0.28 and 0.40 ± 0.15 mg thallium/kg fresh weight, respectively, in muscles, kidneys, livers and testes (Hapke et al., 1980). In pigs fed with 1.45 mg thallium/kg food (dry weight) for 5 months, muscle, kidneys and liver contained 0.18 ± 0.04, 0.44 ± 0.06 and 0.31 ± 0.09 mg thallium/kg fresh weight, respectively. Feeding with 2.71 mg thallium/kg dry weight resulted in 0.39 ± 0.07 (muscle), 0.7 ± 0.2

Table 15. Concentrations in animals from thallium-contaminated areas

Animal	Organ	Source	Locality	n[a]	Concentration of thallium (mg/kg fresh weight)[b]			Reference
					Mean	Range	Highest value	
Fish								
Morone labrax[c]	muscle	-[d]	Taranto, Italy	-	77			Palermo et al. (1983)
Eel[c]	muscle	-	Taranto, Italy	-	96			Palermo et al. (1983)
Salmon	muscle	mining	New Brunswick, Canada	3-4		5.1[e]; 14.6[f]		Zitko et al. (1975)
	liver			3-4		6.8; 23.5		
	gill			3-4		1.2; 30.0		
Salmon	muscle	mining	New Brunswick, Canada	3-4		3.6-34[g]		Zitko et al. (1975)
	liver			3-4		5.7-46		
	gill			3-4		7-89		
Silver-scaled fish	liver	cement plant	Lengerich, Germany	1	1.6			LIS (1980)
	kidney			1	1.3			
	brain			1	0.46			
Mugil cephalus[c]	muscle	-	Taranto, Italy	-	84			Palermo et al. (1983)

Table 15 (contd).

Animal	Organ	Source	Locality	n[a]	Concentration of thallium (mg/kg fresh weight)[b]			Reference
					Mean	Range	Highest value	
Trout	liver	cement plant	Lengerich, Germany	4			0.13	LIS (1980)
	kidney			4			0.885	
	muscle			3			0.09	
Birds								
Duck	liver	industry	Harburg, Germany	30	0.191	< 0.075-0.86		Holm et al.
Duck	kidney	industry	Harburg, Germany	30	0.076	< 0.075-0.43		(1987)
Duck	liver	industry	Stade, Germany	24	0.072			Holm
Duck	liver	industry	Stade, Germany	10	0.186			et al.
Duck	kidney	industry	Stade, Germany	10	0.042			(1987)
Duck	liver	cement plant	Lengerich, Germany	4			0.4	LIS (1980)
	muscle			4			0.4	
Geese[c]	muscle	poison	USA	17	29	4-57		Shaw (1932)
Hen	egg	cement plant	Lengerich, Germany	24	1.26			Kemper & Bertram (1984)
Hen	egg	cement plant	Lengerich, Germany	26			1.6	LIS (1980)
	liver			17			0.8	
	muscle			26			0.8	
	heart			5			0.7	
	stomach			9			0.9	

Table 15 (contd).

Pigeon	liver	cement plant	Lengerich, Germany	1		0.6	LIS (1980)
	kidney			3		0.6	
	muscle			5		0.4	
	heart			1		0.3	
Mammals							
Cattle	kidney	cement plant	Lengerich, Germany	58		2.2	LIS (1980)
	muscle			61		1.5	
Cattle	kidney	not specified	Germany	2		24.0	Frerking et al.
	liver	(presumably		2		2.3	(1990)
	muscle	cement plant)		1		0.4	
	urine			3		1.35[h]	
Fox[c]	liver	poison	Denmark	27		64.0	Munch et al.
	kidney			16		34	(1974)
	intestine			7		55	
Marten[c]	liver	poison	Denmark	15		57	Clausen &
	kidney			16		92	Karlog (1974)
	intestine			9		42	
	brain			5		8.2	

Table 15 (contd).

Animal	Organ	Source	Locality	n[a]	Concentration of thallium (mg/kg fresh weight)[b]			Reference
					Mean	Range	Highest value	
Pig	liver	cement plant	Lengerich, Germany	3			1.2	LIS (1980)
	kidney			296			1.3	
	muscle			300			0.6	
Rabbit	liver	cement plant	Lengerich, Germany	49			5.8	LIS (1980)
	kidney			44			29.0	
Roe deer	liver	cement plant	Lengerich, Germany	1			2.6	LIS (1980)
	kidney			3			14.0	
	muscle			2			1.6	
	heart			3			2.9	
Sheep	liver	cement plant	Lengerich, Germany	4			0.6	LIS (1980)
	kidney			10			1.1	
	muscle			14			1.1	

[a] Number of measurements (animals)
[b] mg/kg fresh weight unless otherwise stated
[c] Fatal poisoning
[d] No data given
[e,f] The two thallium concentrations of 45[e] and 100[f] µg/litre water are in the range of the natural concentrations at that locality; thallium concentration in the gill was higher (25.6 mg/kg fresh weight) at a lower concentration in water (17.9 µg/litre)
[g] Lethal dose of 100 to 10 000 µg/litre
[h] mg/litre

(kidney) and 0.53 ± 0.1 (liver) mg thallium/kg fresh weight. Since 0.5 mg/kg fresh weight is the limit set by the federal state of North Rhine-Westphalia for thallium concentrations in human food, a critical level for pigs seems to be a daily intake corresponding to 1.9 mg thallium/kg dry matter of food (Konermann et al., 1982).

Exposure of farm animals to thallium in the vicinity of the cement plant in Lengerich, Germany, resulted in increased thallium levels in the liver and kidneys of various animals (LIS, 1980): 0.8% of the samples of internal organs contained > 10 mg/kg fresh weight, 1.3% contained 5 to 10 mg/kg, 12.6% contained 1 to 5 mg/kg and 14.5% contained 0.5 to 1 mg/kg. In 0.2%, 3% and 4.4% of meat from various farm animals, 5-10, 1-5 and 0.5-1 mg/kg were found, respectively. Concentrations above 0.5 mg/kg fresh weight were sometimes also found in eggs and chicken meat (up to 0.8 mg/kg), rabbit meat (up to 5.8 mg/kg) and roe deer (1.6 mg/kg) (Table 15) (LIS, 1980). In whole eggs with a concentration of 1.26 mg/kg fresh weight, the concentration in albumin and yolk was 0.394 mg/kg, while in the shell it was 4.94 mg/kg (Kemper & Bertram, 1984).

5.2 General population exposure

The US Environmental Protection Agency calculated the typical value for the exposure of the general population to be 0.48 ng thallium/m^3 air (US EPA, 1980). US EPA (1980) calculated an absorbed amount of 3.4 ng/day assuming an inspired volume of 20 m^3/day and 35% deposition in the lungs. BGA (1979) calculated the daily uptake via the respiratory system to be < 5 ng thallium per day.

More than 99% of samples of drinking-water in the USA contained no thallium (detection limit, 0.3 μg/litre), and the positive samples contained about 0.89 μg/litre. With a water consumption of 2 litre/day, this would result in an intake of < 1 μg thallium/day for most adults (US EPA, 1980). The thallium concentration in 17 bottled mineral waters ranged between < 0.6 and 3.5 μg/litre, but only 4 contained > 2 μg/litre (Korkisch & Steffan, 1979).

In the United Kingdom, Sherlock & Smart (1986) reported the total dietary intake of thallium, based on the analysis of 13 diets. Four (meat, fish, fats and green vegetables) out of nine food groups contained samples with concentrations of thallium above the limit of determination, ranging from 10 to 50 μg/kg fresh

Table 16. Bioconcentration factor[a] of thallium in broilers and laying hens[b] and half-life in the tissues of laying hens

	Bioconcentration factor		
Organ	Broiler	Laying hen	Half-life (days)
Bone	0.54	0.26	2.0
Egg yolk	-	0.26	4.1
Egg albumen	-	0.14	1.6
Egg shell	-	3.72	2.5
Fat	0.006	0.001	-
Feather	0.074	0.006	-
Kidney	0.77	0.38	3.6
Liver	0.19[c] 0.11[d]	0.1	4.0
Muscle	0.46	0.18	3.8
Skin	0.2	0.08	-

[a] Concentration of thallium in the respective tissue divided by the concentration of thallium in the food
[b] From: Ueberschär et al. (1986)
[c] After 3 weeks
[d] After 6 weeks

weight depending on the food commodity. The average dietary intake of thallium for adults was estimated to be 0.005 mg/day with a range of 0-0.01 mg/day, assuming that concentrations less than the limit of determination were equal to zero. The daily intake of thallium from vegetables alone is estimated to be about 3.8 μg for an average adult in the USA (US EPA, 1980). Food of plant origin often contains more thallium than food of animal origin, a 4-fold higher concentration of thallium being eliminated in the urine of vegetarians than in that of humans eating food of varying origin (Ohnesorge, 1985).

A minor route of thallium uptake can be sodium-free dietetic salt (KCl), which contains up to 420 μg thallium/kg salt (Toots & Parker, 1977), but not sodium chloride, which only contains 0.08 μg/kg (Geilmann et al., 1960). Wine has also been found to contain small amounts of thallium (0.056 to 0.684 μg/litre) (Geilmann et al., 1960).

5.3 Occupational exposure during manufacture, formulation or use

Few data on occupational exposure to thallium are available. An industrial plant in the USA used concentrated thallium salt solutions for separations by centrifugation. Considerable variations in the thallium content of the air occurred during the day, depending on the emission potential of the different steps of the procedure (Hill & Murphy, 1959), although no exact data were reported. In a plant in the United Kingdom manufacturing special alloy anodes for use in magnesium seawater batteries, air samples from two working areas contained a maximum level of 0.022 mg thallium/m^3 and 0.014 mg/m^3 (Marcus, 1985). Detailed data are available from a Russian plant producing thallium before and after changes in processes. The air concentration varied from 0.12 to 0.18 mg thallium/m^3, but peak concentrations of 13.5 to 17.4 mg/m^3 during the smelting process were observed. During dissolving and packing of thallium salts, the air thallium concentrations were 0.117 and 0.274 mg/m^3, respectively. After changes in the smelting process were instituted, the air thallium content decreased to 0.0036-0.0072 mg/m^3 (Tikhova, 1964). Floating dust in a German thallium smelter contained 60 to 9700 mg thallium/kg dust; the air-suspended dust concentration was 6-50 μg/m^3 (Briese et al., 1985).

Working with thallium causes dust contamination of the hands, this increasing with the duration of work (Tikhova, 1964). The dust concentration on the hands has been reported to be in the range of 0.04 to 10.6 mg/m^2 (Shabalina & Spiridonova, 1979).

6. KINETICS AND METABOLISM

Investigations into the kinetics and metabolism of thallium in aquatic and terrestrial animals have mainly made use of radioactive compounds, especially thallium-201. The investigations cited in this chapter have been performed with various thallium salts, but to facilitate comparison concentrations have been generally expressed as μg (or mg) thallium/litre.

6.1 Absorption

6.1.1 Animals

6.1.1.1 Aquatic animals

Clams and mussels reached equilibrium within 12 to 19 days when exposed to 50 or 100 μg thallium/litre. Depending on the exposure level, the clams contained 5 or 9 mg/kg dry weight and the mussels 3 or 5 mg/kg (Zitko & Carson, 1975).

Very poor absorption of thallium salts from water containing 0.1 mg thallium/litre was found in isolated gill preparations of the mussel *Mytilus galloprovincialis* (Nolan et al., 1984). In the first 10 min about 10% of the dose was absorbed.

Accumulation of thallium in juvenile salmon exposed for 300 h to different concentrations of thallium (17.9 to 200 μg thallium/litre) varied in different organs (Table 15). In muscle, thallium levels increased almost linearly with the water thallium concentration from 2.3 to 27.0 mg/kg tissue (wet weight). Data on the liver show no consistent trend, but in gills an obvious maximal accumulation capacity of about 30 mg/kg was reached even at the lowest concentration of 17.9 μg/litre water (Zitko et al., 1975). The mean accumulation factor (mg thallium per kg tissue wet weight divided by mg thallium per litre water) of the gills (up to 1430) is about three to ten times higher than that of muscle or liver.

6.1.1.2 Terrestrial animals

Using different routes of administration of thallium(I) nitrate solution (oral, intratracheal, subcutaneous, intraperitoneal, intramuscular and intravenous), thallium was rapidly and almost completely absorbed in rats (Lie et al., 1960). High concentrations of thallium were detected in the blood within just 1-2 h after oral administration of thallium(I) malonate and thallium(I) sulfate (Aoyama, 1989), and as little as 1 h after oral or parenteral administration of thallium(I) sulfate it was found in the urine and faeces of rats (Lund, 1956a).

Only a few experimental studies on intestinal absorption are available. In sheep and cows about 2% of the thallium ingested with contaminated food was retained, while about 98% was eliminated (Crössmann, 1984). Sabbioni et al. (1980b) found no obvious differences using various doses of thallium(I) or thallium(III) sulfate or bromide, but, after oral uptake of dimethyl thallium(III) bromide in rats, the organs contained only 1 to 10% of the concentrations found after the uptake of inorganic thallium, indicating reduced absorption. In rats the absorptive capacity of different ligated regions of the intestinal tract varies strongly: ^{201}thallium (sulfate) was rapidly absorbed from the colon, but more slowly from the ileum and jejunum and slowest from the stomach (Sabbioni et al., 1984a). Within 60 min the ligated colon absorbed about 75%. Lower values were obtained for the other regions. Voltage clamp experiments on the mucosa of rat descending colon showed exclusive transport by diffusion to the serosal side (Schäfer et al., 1981).

Absorption of thallium through the skin of rats is indicated by the determination of a cutaneous LD_{50} of 117 mg/kg for thallium(I) carbonate (Shabalina et al., 1980).

6.1.2 Humans

Increased levels of thallium have been observed in the lungs of coal miners, but no data are available concerning the absorption of thallium salts after inhalation exposure (section 6.2.2.2) (Weinig & Zink, 1967). Generally it is assumed that about 35% of respirable dust is deposited in the lung (Ohnesorge, 1985; IPCS, 1992) and that up to 100% of the deposited thallium is absorbed (Gubernator et al., 1979). The rate of deposition and absorption is high because thallium concentrations increase markedly with decreasing particle size (sections 3.2.3.2 and 5.1.1.2), and small particles become deposited in the lung whereas larger particles are deposited in the upper respiratory system (Natusch & Wallace, 1974). In addition, nearly all the thallium chloride in the dust emitted from the cement plant in Lengerich was water-soluble (LIS, 1980). In human broncho-alveolar lavage fluids, 0.258 ng thallium per 1000 cells was found in a silicosis patient, but only

0.009-0.05 ng in patients suffering from other lung diseases (Maier et al., 1986).

From several intoxication cases, e.g., after oral and topical application of thallium(I) sulfate during depilatory treatment (Barckow & Jenss, 1976; Schmidbauer & Klingler, 1979), it can be assumed that both percutaneous and gastrointestinal absorption occur, but no data on absorption are available. High blood thallium concentrations have been reported following human poisoning (see Table 20).

6.2 Distribution

6.2.1 Animals

No effect of the route of administration (oral, intratracheal, subcutaneous, intraperitoneal, intramuscular and intravenous) on distribution was observed by Lie et al. (1960). After intravenous injection, an initial increase in the thallium concentration of the blood is followed by a steep decrease within 5 to 15 min (Gehring & Hammond, 1967; Lameijer & van Zwieten, 1977a,b). A similar trend is observed when concentrations are compared 1.5 and 24 h after oral administration of 10 mg thallium/kg body weight to rats or intravenous injection into rabbits (Careaga-Olivares & Morales-Aguilera, 1993). Thallium is distributed by the blood stream to all organs.

Data on the distribution of thallium in the main blood compartments, e.g., serum and erythrocytes, have been reported. *In vitro* measurements by Lund (1956a) and Witschi (1965) indicate that thallium is distributed equally in blood plasma and red cells, presumably without any direct binding, while Gregus & Klaassen (1986) found that 2 h after intravenous injection of 1 to 30 mg/kg more than 80% of blood thallium was in the plasma. According to Ducket et al. (1983), 24 h after intraperitoneal injection of a very low dose of thallium, nearly 90% was located in the red blood cells and only about 10% in the plasma. An intermediate result was obtained by Ulrich & Long (1955): after intraperitoneal injection of about 20 μg radioactive thallium, about one third of the thallium concentration in the whole blood was located in the plasma. This ratio did not change over the period 0.5 h to 96 h after injection, although the concentration in the blood decreased with time. Similar results were obtained in human whole blood (*in vitro*) and in rabbit whole blood (*in vitro* and *in vivo*) (Careaga-Olivares & Morales-Aguilera, 1990). A slightly higher concentration in the erythrocytes was also found by Leloux et al. (1987), but during redistribution phases the concentrations in erythrocytes were increased, suggesting that erythrocytes may be involved in thallium uptake.

6.2.1.1 *Distribution after administration of a single dose*

The distribution of thallium after administration of single doses of thallium compounds has been investigated in a number of studies, using either subtoxic doses (0.2 µg/kg to 8 mg thallium/kg body weight) (e.g., Ulrich & Long, 1955; Lie et al., 1960; Bradley-Moore et al., 1975; Edel Rade et al., 1982; Ziskoven et al., 1983; Ducket et al., 1983) or toxic doses (e.g., Lund, 1956a; Fitzek & Henning, 1976; Achenbach et al., 1980; Leloux et al., 1987b; Aoyama, 1989). Comparisons of dose-dependent distribution (Emara & Soliman, 1950; Gehring & Hammond, 1967; Sabbioni et al., 1980a,b, 1982; Talas & Wellhöner, 1983; Gregus & Klaassen, 1986; Ríos et al., 1989) revealed only slight differences in the distribution pattern, even at intraperitoneally administered concentrations as different as 0.00004, 2, 20 and 2000 µg thallium/rat (Sabbioni et al., 1980a).

The distribution of thallium in the organs is a time-dependent process (Tables 17 and 18). Summarizing the investigations on rats, rabbits, dogs and goats, in an initial phase after a single dose of an inorganic thallium compound, e.g., thallium sulfate or thallium chloride, maximal concentrations occurred in the kidneys and nearly equal levels were found in the testes, myocardium, salivary glands, muscle, liver, intestines, adrenals and thyroid; fat and brain contained very low levels of thallium (Lund, 1956a; Gehring & Hammond, 1967; Bradley-Moore et al., 1975; Sabbioni et al., 1982; Talas & Wellhöner, 1983). About 24 h after administration, the relative thallium content of all organs, with the exception of the kidneys, decreased and that of brain, muscle and testes increased. In guinea-pigs administered lethal doses of thallium, kidney and liver finally contained about equal levels, presumably due to kidney damage (Weinig & Walz, 1971). With respect to the affinity and redistribution of thallium in rat organs, Leloux et al. (1987b) distinguished three compartments according to affinity and redistribution which are not completely in agreement with the experimental data shown in Table 17.

Intraperitoneal application of 16, 32 and 48 mg thallium sulfate/kg body weight to male rats resulted in peak levels occurring in various regions of the brain 24 h after injection,

Table 17. Alterations in the distribution of thallium in different organs of experimental animals at different times after administration (mean ± standard deviation)[a]

	Rats (4; intraperitoneal)[b]				Syrian hamsters (5; oral)[c]			
Experiment No.:	1		2		3		4	
	2 h	40 h	4 h	24 h	1 h	24 h	1 h	24 h
	(% of dose)		(% of dose of thallium/g wet weight)		(mg thallium/kg wet weight)			
Blood	0.05 ± 0.01[d]	0.08 ± 0.05[d]	0.023 ± 0.001	0.027 ± 0.017	1.5 ± 0.6	1.0 ± 0.1	1.7 ± 0.6	0.8 ± 0.1
Bone	-	-	0.239 ± 0.036	0.216 ± 0.064	-	-	-	-
Brain	0.027 ± 0.003	0.28 ± 0.05	0.047 ± 0.009	0.090 ± 0.023	< 0.1	3.7 ± 0.3	0.6 ± 0.1	3.0 ± 0.3
Heart	0.57 ± 0.07	0.33 ± 0.03	0.546 ± 0.064	0.337 ± 0.006	11.4 ± 1.7	7.2 ± 0.7	21.0 ± 3.4	6.6 ± 0.7
Intestine	1.1 ± 0.26	-	-	-	-	-	-	-
Kidney	5.65 ± 0.35	9.75 ± 0.97	3.354 ± 0.535	1.899 ± 0.253	58.5 ± 8.4	51.3 ± 14.0	88.4 ± 21.8	41.5 ± 0.9

Table 17 (contd).

Liver	4.44 ± 0.55	0.95 ± 0.19	0.373 ± 0.040	0.228 ± 0.028	14.3 ± 7.6	6.6 ± 2.1	39.7 ± 6.2	5.8 ± 0.4
Lung	0.55 ± 0.35	0.45 ± 0.04	0.289 ± 0.076	0.230 ± 0.081	-	-	-	-
Muscle	0.6 ± 0.02[e]	0.54 ± 0.2[e]	0.168 ± 0.031	0.234 ± 0.091	< 0.1	9.1 ± 2.4	1.2 ± 0.5	9.1 ± 1.2
Pancreas	0.86 ± 0.34	0.57 ± 0.13	0.505 ± 0.163	0.310 ± 0.206	-	-	-	-
Salivary gland	0.76 ± 0.06	0.7 ± 0.32	0.595 ± 0.078	0.370 ± 0.090	-	-	-	-
Spleen	0.32 ± 0.03	0.18 ± 0.02	0.338 ± 0.039	0.249 ± 0.054	-	-	-	-
Testes/ Ovary	0.56 ± 0.04	0.93 ± 0.34	0.388 ± 0.077	0.200 ± 0.010	< 0.1	14.7 ± 0.7	1.0 ± 0.1	14.0 ± 1.0

[a] - = no data given
[b] No. 1: Sabbioni et al. (1980b) and No. 2: Edel Rade et al. (1982): injection of thallium(I) sulfate (2 μg thallium/rat)
[c] Nos. 3 and 4: Aoyama (1989): administration of 12.5 mg thallium(I) sulfate/kg (No. 3) or 12.35 mg thallium(I) malonate/kg (No. 4)
[d] % dose/ml
[e] % dose/g wet weight

Table 18. Distribution of thallium in different organs of experimental animals after different periods of exposure[a]

Experiment no:[b]	1a	1b	2	3	4a	4b	5
	72 h	72 h	4 h	24 h	48 h	48 h	48 h
Blood	0.007	0.001	0.02	0.02	0.02	0.02	-
Bone	0.016	0.028	0.37	0.42	0.23[c]	0.23[c]	0.67
Brain	0.006	0.008	0.05	0.13	0.17	0.14	0.32
Fat	0.0008	0.001	-	0	0.01	0.02	-
Heart	0.027	0.030	0.55	0.20	0.34	0.25	0.63
Intestine	0.012[d]	0.017[d]	-	0.51	0.49[e]	0.76[e]	0.50
Kidney	0.205	0.210	3.35	2.15	2.58	0.53	5.36
Liver	0.019	0.026	0.37	0.33	0.19	0.16	0.53
Lung	0.016	0.018	0.29	0.55	0.25	0.24	0.51
Muscle	0.014	0.025	0.17	0.49	0.25[f]	0.27[f]	0.76

Table 18 (contd).

Pancreas	0.010	0.014	0.51	-	0.44	0.33	-
Salivary gland	-	-	0.59	-	0.42	0.44	1.06
Skin	-	-		0.20	-	-	0.32
Spleen	0.016	0.021	0.34	0.20	0.24	0.21	0.51
Testes/ Ovary	0.023	0.024	0.39	0.39	0.53	0.53	0.81

[a] Thallium content in % of dose per kg wet weight; - = no data given

[b] No. 1: Talas & Wellhöner (1983): mean of 2 rabbits after intravenous injection of < 2 μg (No. 1a) or 1.1 mg thallium/kg (No. 1b); No. 2: Sabbioni et al. (1982): mean of 4 rats after intraperitoneal injection of 2 μg/rat; No. 3: Lund (1956a): data of 1 rat after intraperitoneal injection of 10 mg/kg; No. 4: Barclay et al. (1953): mean of 4 rats after intravenous injection of 23 μg thallium nitrate (No. 4a) or mean of 3 rats after intravenous injection of 10.023 mg thallium sulfate (No. 4b); No. 5: Lie et al. (1960): mean of 18 rats after injection of thallium by various routes

[c] Femur

[d] Mean of values of small and large intestine

[e] Lower bowel

[f] Abdominal muscle

except in the hypothalamus where the peak was reached earlier. After 24 h, the regional concentrations decreased in the following order: hypothalamus, midbrain, hippocampus, thalamus, pons, cerebellum, corpus striatum and cerebral cortex. In the hypothalamus, a region with low blood-brain barrier protective mechanisms, thallium concentration was significantly higher than in the corpus striatum, whereas in the cerebral cortex it was significantly lower than in all other regions (Ríos et al., 1989). Whereas weanling rats showed a region-dependant distribution of thallium after a single sublethal intraperitoneal injection (16 mg thallium/kg body weight), thallium concentrations were similar in different brain regions of newborn rats (Galvan-Arzate & Ríos, 1994).

Although the endocrine organs were thought to be involved in the mechanism of toxicity, no accumulation was observed in autoradiographic studies of low-dosed adult mice and rats (Barclay et al., 1953; André et al., 1960). Leloux et al. (1987b) found no obvious deviation from the levels found in other organs 5 days after subacute or acute intoxication with 4 or 20 mg thallium nitrate/kg, respectively. Thyresson (1951), however, reported a high concentration of thallium (96.9 mg/kg wet weight) in the thyroid gland of rats 24 h after administration of 40 mg thallium nitrate/kg body weight. According to Ulrich & Long (1955), treatment of rats with thyrotropic hormone subsequent to the administration of thallium did not affect the uptake of thallium by the thyroid, whereas pretreatment (2 days prior to administration) significantly increased the initial uptake. In addition, the authors reported that adrenals, thyroid and pituitary contained similar concentrations of thallium.

Different modes of application and different thallium compounds hardly affect the distribution pattern, as shown by Sabbioni et al. (1980b) with intravenous or oral administration of thallium(I), thallium(III) and dimethyl thallium(III) in rats. Another investigation showed that after administration of thallium(I) malonate and thallium(I) sulfate the distribution pattern between different organs varied only initially. Later, both patterns were similar (Aoyama, 1989).

6.2.1.2 *Distribution after long-term sublethal administration*

The distribution pattern of thallium in chronically poisoned rats shows a strong similarity to the final pattern after a single dose, there being wide distribution throughout the body. Lameijer

& van Zwieten (1977a,b) determined thallium concentrations in urine, blood and 19 different tissues of rats exposed to thallium (10 mg/litre) in drinking-water for 9 or 24 weeks. No statistically significant differences in the concentration of thallium were observed for any tissues except kidney. The thallium concentration in the renal medulla was about 7 times higher than in the heart, liver, muscle, brain or skin. If rats received 30 mg/litre in drinking-water, they died within 9 to 11 days, but after 7 days this administration had not affected the rapid decline of the thallium blood level following administration of an additional intravenous dose of 1 mg/kg (Lameijer & van Zwieten, 1977a).

6.2.1.3 *Transplacental transfer of thallium*

Mice were gavaged with thallium(I) sulfate (8 mg thallium/kg body weight) on gestation day 9 and tissue concentrations were determined 0.5 to 24 h later. Thallium levels in fetuses and maternal kidneys rose during the first hour, then levelled off to a plateau which did not change during the following 22 h. No indication of a specific placental barrier was detected (Ziskoven et al., 1980). Ziskoven et al. (1983) repeated the study, additionally including rats. After 10, 20 and 30 min, increasing thallium concentrations were found in the maternal kidneys, reaching a plateau after 30 to 60 min, which was stable for at least 50 h (last measurement). A similar time course of thallium uptake was observed in the fetal tissues: the initial uptake was comparable to that of the maternal kidney, but the resultant concentrations were 10-fold lower. There were no specific differences between mice and rats.

A slightly delayed transfer to fetuses was observed in an autoradiographic study with mice (gestation day 15), dosed with an intraperitoneal injection of thallium sulfate. After just 15 min, thallium was observed in the fetuses, but the maximum level was reached within 2 to 4 h, when some thallium elimination took effect. During the whole observation period, fetal thallium levels were lower than those of the placenta (Olsen & Jonsen, 1982). The authors also reported on the influence of the stage of gestation. Thallium crossed the placenta throughout gestation; during the early stages it was concentrated in the visceral yolk sac placenta and during late gestation additionally in the chorioallantoic placenta and amnion.

When near-term mice and rats (gestation day 17-18) were given subcutaneous injections of ^{204}thallium sulfate, thallium concentrations in the mouse fetuses rose during the following 8 h and in the rat fetuses during the following 16 h. Thereafter, the fetal/maternal ratios in tissues remained constant at 0.84 in rats and considerably lower (0.46) in mice (Gibson et al., 1967).

Intra-arterial infusion of 0.2 to 6.4 mg ^{204}thallium sulfate/min per kg body weight on day 20 of pregnancy in rats resulted in an initially restricted transplacental transfer. In the lowest and highest dosage groups, 32 min after administration, thallium levels in fetal tissues corresponded to only about 7% of those in the maternal plasma, perhaps because two-thirds of the whole blood thallium was located in the erythrocytes and did not pass the placental barrier (Gibson & Becker, 1970).

Sabbioni et al. (1982) compared placental transfer in rats after intraperitoneal injection of a low dose (2 μg thallium-201/rat; gestation day 13) with that after administration of a toxic dose by gavage (10 mg thallium/kg body weight; gestation day 16). Thallium concentrations in maternal and fetal brain were similar 4 h after injection of the low dose. In fetal liver they were 80% lower than in the maternal liver, and in the placenta and fetal organs they were higher than in the blood of the dams and fetuses. Very low concentrations were found in the amniotic fluid. After 8 days, concentrations in most of the maternal organs were about 10 times lower than the initial levels, while in muscle, cerebellum and brain of the dams, reductions of only 60, 40 and 5%, respectively, were found during the same period. In contrast to the variable decline in maternal organs, a 10-fold decline was observed in the whole fetus and its liver, brain and blood. Preliminary results of the authors indicated that this was due to a much stronger thallium accumulation in the mitochondria of the adult rat brain than in those of the fetal brain. Furthermore, identical low-dose experiments (2 μg/rat) also showed a faster decline of thallium concentrations in fetal than in maternal brain, after an initially more rapid entry into the fetal brain, whereas declines in the liver levels were nearly identical (Edel Rade et al., 1982). Administering a toxic dose of 10 mg/kg to the dams resulted in 60% lethality within the following 3 days. After 3 days, concentrations in the liver and brain of the surviving dams were similar and about 2-fold higher than in the corresponding fetal organs. Therapeutic oral dosing with Prussian Blue (100 mg/kg body weight, twice daily) starting 8 h after administration of thallium, significantly reduced the thallium concentrations in maternal and fetal tissues and only 1 of 12 adult rats died (Sabbioni et al., 1982).

Transplacental transfer of thallium has also been observed in a cat (Fitzek & Henning, 1976). The cat showed signs of a strong thallium intoxication and was killed after abortion of approximately 5-week-old fetuses. Thallium levels in maternal blood and fetal tissues were similar, but the concentrations in fetal heart and lungs were two to three times higher than in the corresponding maternal organs.

6.2.2 Humans

Background thallium concentrations found in human body fluids and tissues are given in Table 19. After poisoning, thallium concentrations ranging up to nearly 36 mg/litre in blood, 25 mg/litre in urine and 8 mg/kg in hair have been found (Table 20).

6.2.2.1 Increased concentrations after lethal poisoning

In reports of postmortem examinations after suicide or homicide, data on the distribution of thallium in different organs are rarely included with data on dose and application routes (Table 20). The distribution pattern shows no consistent trend. In a single individual, concentrations in bones, fat and muscles from different parts of the body may vary, e.g., in vertebrae (12.7 mg/kg), sternum (7.0 mg/kg), femur (16.4 mg/kg) and tibia (9.0 mg/kg) (case 7 in Table 21) (Arnold, 1986). The distribution of thallium differs considerably from that reported for potassium in humans (Davis et al., 1981). Endocrine glands, kidneys, liver and intestine (without content) showed the highest concentrations (Table 21).

With respect to the total amount per organ, liver or lung were found to contain 2 to 6 times and the brain about 1.5 to 2 times more thallium than the kidneys (Curry et al., 1969; Arnold, 1986).

A comparison between the white and grey matter of the brain revealed that in the latter the concentration was three times higher (Cavanagh et al., 1974). Detailed data of thallium concentrations in different regions of the nervous system were given by Davis et al. (1981). The authors showed that areas of the brain rich in neurons tend to accumulate twice as much thallium as areas devoid of neurons, and that the grey matter contains some of the highest thallium levels of any body organ.

Table 19. Background concentrations of thallium in humans

Material	Number of measurements	Concentration of thallium		Concentration unit	Reference[f]
		Mean ± SD	Range		
Blood, whole	2		0.33-0.59	μg/litre	Weinig & Zink (1967)
			< 20		Bowen (1966)
	13		0.47-9		Iyengar et al. (1978)
	320		< 5-80		Singh et al. (1975)
			0.05		Kemper (1979)
	418	0.39 ± 0.05	0.1-1.1		Minoia et al. (1990)
			0.5-2		Kemper & Bertram (1991)
plasma			< 2.5		Bowen (1966)
	1		< 2.5		Iyengar et al. (1978)
Bone	2		0.84-2.51	μg/kg fresh weight	Weinig & Zink (1967)
	1		2		Iyengar et al. (1978)
	5		< 0.1-0.1		Goenechea & Sellier (1967)
Bone[a]	1		0.7; 0.9		Goenechea & Sellier (1967)
Brain			< 0.5	mg/kg dry weight	Bowen (1966)

Table 19 (contd).

Bronchoalveolar lavage fluids	1[b]		0.258	ng/1000 cells	Maier et al. (1986)
	1[c]		0.009		
	1[c]		0.011		
	1[d]		0.016		
	1[d]		0.050		
Faeces	5		< 0.02-3.0	μg/kg fresh weight	Goenechea & Sellier (1967)
Hair	7	18.6 ± 14.9	7-51	μg/kg fresh weight	Geilmann et al. (1960)
	6	10.4 ± 4.3	4.8-15.8		Weinig & Zink (1967)
	1		< 20		Ziegler & Ziegler (1984)
Heart			< 0.4	mg/kg dry weight	Bowen (1966)
Kidney			< 0.4	mg/kg dry weight	Bowen (1966)
Kidney	6	2.7 ± 1.1	1.44-4.1	μg/kg fresh weight	Weinig & Zink (1967)
	8		< 3		Iyengar et al. (1978)
	259		0.03-8.6		Bösche & Magureanu (1983)
Liver			0.4	mg/kg dry weight	Bowen (1966)
	11	0.47 ± 0.13	< 0.4-0.9		Johnson (1976)

Table 19 (contd).

Material	Number of measurements	Concentration of thallium		Concentration unit	Reference[f]
		Mean ± SD	Range		
Liver	6	1.1 ± 0.9	0.55-2.85	µg/kg fresh weight	Weinig & Zink (1967)
	1		0.4		Goenechea & Sellier (1967)
	6		1-9		Iyengar et al. (1978)
			0.5-3		Kemper & Bertram (1991)
Lung			< 0.3	mg/kg dry weight	Bowen (1966)
Lung	4	1.1 ± 0.7	0.36-1.8	µg/kg fresh weight	Weinig & Zink (1967)
Muscle			< 0.4	mg/kg dry weight	Bowen (1966)
Muscle	6	2.1 ± 2.1	0.52-7.05	µg/kg fresh weight	Weinig & Zink (1967)
	3		15-100		Iyengar et al. (1978)
Muscle[e]	1		0.4	µg/kg fresh weight	Goenechea & Sellier (1967)
Nail	6	51.2 ± 12.1	40-74	µg/kg fresh weight	Geilmann et al. (1960)
	6	2.6 ± 1.4	0.72-4.93		Weinig & Zink (1967)
Skin			< 0.2	mg/kg dry weight	Bowen (1966)

Table 19 (contd).

Urine	10		< 0.02-1.0	μg/kg fresh weight	Goenechea & Sellier (1967)
Urine			0.05-0.1	μg/litre	Geilmann et al. (1960)
	14	0.7 ± 0.5	0.07-1.69		Weinig & Zink (1967)
			0.05-20		Kemper (1979)
	31	0.4 ± 0.2	< 0.1-1.2		Brockhaus et al. (1981b)
	10	0.3 ± 0.2	< 0.1-0.9		Brockhaus et al. (1981b)
	149	0.3 ± 0.14	0.02-0.7		Dolgner et al. (1983)
	72	0.22 ± 0.14	0.06-0.61		Apostoli et al. (1988)
	496	0.42 ± 0.09	0.06-0.82		Minoia et al. (1990)
			0.05-1.5		Kemper & Bertram (1991)
Urine	20		< 0.3-1.1	mg/kg creatinine	Schaller et al. (1980)
	10	2.2 ± 1.6			Briese et al. (1985)

[a] 1.5 months after death
[b] silicosis patient
[c] saw setters suffering pneumoconiosis
[d] welders suffering emphysema
[e] 6 months after death
[f] Additional values for other tissues have been compiled by Iyengar et al. (1978)

Table 20. Concentrations of thallium in cases of poisoning

Material	Number of cases	Range of thallium concentrations	Concentration unit	Reference
Blood, whole		50-6000[a]	μg/litre	Kemper (1979)
Blood	2	29; 7700		Alarcón-Segovia et al. (1989)
Blood, whole	3	350-36 000		Heath et al. (1983)
Blood, plasma	2	300; 1500		Heath et al. (1983)
Blood, erythrocyte	2	400; 2300		Heath et al. (1983)
Bone	1[b]	0.9-2.1	μg/kg fresh weight	Goenechea & Sellier (1967)
Faeces	1	6500-38 400		Paulson et al. (1972)
Hair	1	650		Geilmann et al. (1960)
Hair	1[b]	6.8		Goenechea & Sellier (1967)
Hair	1	420-1800		Hagedorn-Götz & Stoeppler (1975)
Hair		250-8000[a]		Kemper (1979); Kemper & Bertram (1984)
Heart	1	3600		Munch et al. (1933)
Intestine	1	3600		Munch et al. (1933)
Intestine	1	0.8; 4.0		Goenechea & Sellier (1967)

Table 20 (contd).

Kidney	5	2700-11 600		Munch et al. (1933)
Kidney	1	106 000		Heath et al. (1983)
Liver	1	75 000		Heath et al. (1983)
Liver	2	3700; 5500		Munch et al. (1933)
Lung	2	3300; 7700		Munch et al. (1933)
Muscle	1[b]	< 0.02; 1.3		Goenechea & Sellier (1967)
Nails	1	2400		Geilmann et al. (1960)
Spleen	3	2900-6600		Munch et al. (1933)
Urine		50-25 000[a]	μg/litre	Kemper (1979)
Urine	15	500-20 400		Klöppel & Weiler (1978)
Urine	1	3100		Gastel (1978)
Urine	3	10-13 800		Alarcón-Segovia et al. (1989)
Urine	2	2700-30 000	μg/litre fresh weight	Heath et al. (1983)
Urine	1	0.65	mg/kg creatinine	Hagedorn-Götz & Stoeppler (1975)

[a] Concentrations indicative for poisoning
[b] 3 years after death

Table 21. Concentrations of thallium in individual cases of human poisoning

	Thallium concentration (mg/kg wet weight or mg/litre)						
Case no:[a]	1	2	3	4	5	6	7
Duration[b]:	> 14 days	> 21 days	9 days	8 days	11 days	12 days	13 days
Dose:	-[c]	-	5-10 g	0.75 g[d]	-	-	-
Adrenal	-	-	83.6	-	-	-	-
Blood	5.1	3.4	5.1	-	-	3.0	-
Bone	-	5.0	-	0.92	1.9	8.0	7.0-16.4
Brain	8.5	-	62-140	0.15	-	-	-
grey	-	10.0	-	-	-	-	-
white	-	3.0	66.1	-	-	-	-
cerebellum	-	-	103.3[e]	-	1.5	5.0	-
cerebrum	-	-	102.0[e]	-	1.0	-	-
Fat	-	-	< 1.0	-	0.4-1.2	-	-
Heart	-	13.3	-	0.19	1.5	13.0	6.2
left	-	-	26.8	-	-	-	-
right	-	-	131.6	-	-	-	-
Intestine	-	-	-	0.20	-	-	-
small	4.4[f]	-	-	-	0.5-0.9	8.0[f]	6.4
colon	71.0[f]	120.0[f]	126.0	-	2.0	500.0[f]	8.5

Table 21 (contd).

Kidney	26.7	20.0	74.1	0.26	3.0	28.0	12.5
Liver	8.6	5.0	77.3	0.82	1.8	15.0	14.7
Lung	-	1.8	-	0.15	0.9	4.0	-
Muscle	10.1	5.0	26.8	0.21	0.4-2.0	-	3.6
Pancreas	-	-	71.7	-	-	-	-
Parathyroid	-	-	38.1	-	-	-	-
Pituitary	-	-	114.5	-	-	-	-
Salivary gland	-	-	32.1	-	-	-	-
Skin	-	6.0	32.1	-	0.3	-	-
Spleen	-	-	-	0.35	1.9	-	-
Testes/Ovaries	-	-	152.0	-	-	-	-
Thyroid	-	-	33.5	-	4.6	-	-
Urine	15.6	5.9	3.3	-	-	3.0	-

[a] Case no. 1: Curry et al. (1969); no. 2: Cavanagh et al. (1974); no. 3: Davis et al. (1981); no. 4: Graben et al. (1980); nos. 5-7: Arnold (1986); nos. 3-7: poisoning by one uptake of thallium

[b] Period of time from uptake of thallium to death or determination of concentrations

[c] - = no data given

[d] 15-week-old embryo; dose of mother

[e] Cortex

[f] Content

6.2.2.2 *Increased concentrations after long-term sublethal poisoning*

Thallium levels in urine (Table 22), blood or saliva of chronically exposed people offer better indications of the actual burden than those derived from hair samples, since elevated levels in hair can be caused by exogenous dust (Bertram et al., 1985).

People consuming food grown in private gardens and living at a distance of more than 3 km from the cement plant at Lengerich, Germany showed significantly higher concentrations of thallium in their urine, decreasing with increasing distance from the plant, than people who did not consume food from their gardens. Thallium concentrations in the urine of people living near the plant (< 1 km) and consuming food grown in private gardens were about five times higher (3.95 μg/litre) (Brockhaus et al., 1980). Peak values were 76.5 μg/litre in urine and 565 μg/kg in hair (Ewers & Brockhaus, 1982). In this area a medical survey was carried out immediately after the occurrence of thallium emissions had been recognized; urine thallium levels in about 80% of the population were found to exceed the upper normal limit of 1 μg/litre (Brockhaus et al., 1980; Dolgner et al., 1983). The recommendation to avoid home-grown vegetables was followed by many people and resulted in a significant decrease in urine thallium levels. However, in some residents, even 8 years later, increased levels of > 20 μg/litre urine could be found. Probably there was still a significant contamination of soil and thus of home-grown vegetables (Ewers, 1988).

Subsequent studies were carried out at other cement factories. About 70% of employees at two cement plants in Middle and Lower Franconia, Germany were found to have normal thallium concentrations in their urine. However, at a third factory in the same area only 30% of employees showed normal thallium urine levels, presumably because of the higher thallium content in the raw material used (Schaller et al., 1980). The population around the three cement plants showed normal urine thallium levels: of 238 people tested, 194 had thallium concentrations below 2 μg/litre, 36 were in the range of 2 to 5 μg/litre, and 5 were between 5 and 10 μg/litre. Higher concentrations were found in the urine of three people (11.5, 14.5 and 19.5 μg/litre) (Steuer, 1980).

Only a few cases resulting from industrial exposure have been reported and seem to be mainly a result of skin contact or inhalation (Kazantzis, 1986; Ewers, 1988).

Table 22. Concentrations of thallium following environmental or occupational exposure

Material	Number of people	Concentration of thallium		Concentration unit	Source of thallium	Reference
		Mean ± SD (Median)	Range			
Hair	1163	20.3 ± 42.7	0.6-565	µg/kg fresh weight	cement plant	Brockhaus et al. (1981b)
Urine	50	(0.6)	< 0.3-4.9	mg/kg creatinine	cement plant	Schaller et al. (1980)
	47	(1.65)	0.4-6.3	mg/kg creatinine	cement plant	Schaller et al. (1980)
	21	(0.34)	< 0.3-2.9	mg/kg creatinine	cement plant	Schaller et al. (1980)
	10	7.1 ± 6.0		mg/kg creatinine	zinc smelter	Briese et al. (1985)
	1265	5.2 ± 8.3	< 0.1-76.5	µg/litre	cement plant	Brockhaus et al. (1981b)
	82	2.4 ± 4.3	< 0.1-35.8	µg/litre	cement plant	Dolgner et al. (1983)
	117	3.0 ± 5.6	0.2-37.7	µg/litre	cement plant	Dolgner et al. (1983)
	34[a]	3.4 ± 3.5	0.4-14.8	µg/litre	cement plant	Dolgner et al. (1983)
	30	0.38 ± 0.30	0.08-1.22	µg/litre	cement plant	Apostoli et al. (1988)

Table 22 (contd).

Material	Number of people	Concentration of thallium		Concentration unit	Source of thallium	Reference
		Mean ± SD (Median)	Range			
Urine (*contd*)						
	20	0.40 ± 0.34	0.08-1.22	µg/litre	cement plant	Apostoli et al. (1988)
	10	0.33 ± 0.16	0.09-0.60	µg/litre	cement plant	Apostoli et al. (1988)
	9	0.38 ± 0.29	0.10-1.04	µg/litre	iron smelter	Apostoli et al. (1988)
	74[b]	16.0 ± 16.9	0.2-76.5	µg/litre	cement plant	Brockhaus et al. (1981a)
	74	7.9 ± 8.8	0.2-42.6	µg/litre	cement plant	Dolgner et al. (1983)
	21	0.33 ± 0.27	0.06-1.04	µg/litre	iron smelter	Apostoli et al. (1988)
	12	0.29 ± 0.21	0.06-0.70	µg/litre	iron smelter	Apostoli et al. (1988)

[a] children

[b] Data of people with high concentrations of thallium or possibly thallium-related disorders determined in the first survey and about 1 year later in the following line

At a zinc smelter in eastern Germany, increased thallium levels were not only found in the urine of men working in the production process, but also in men working in the administration. During the production of thallium in this plant, the levels were further increased (maximal value: 28.6 μg/litre) (Briese et al., 1985).

High concentrations of thallium were found in lung tissue from two coal miners (20.2 and 29.5 μg/kg wet weight). Concentrations in most other tissues were normal (Weinig & Zink, 1967).

In Italy, slight but significant increases in thallium levels were found in the urine of cement workers (0.4 μg/litre) and cast iron workers (0.3 μg/litre), compared with a non-exposed group (0.2 μg/litre). There was no correlation with age or the duration of exposure (Apostoli et al., 1988).

Weinig & Zink (1967) reported a slight elevation of urine thallium levels of vegetarians (and smokers) compared to controls. However, it should be noted that each group comprised only three people and the levels were far below those of thallium-affected people. Geilmann et al. (1960) estimated that more than 60% of the thallium content of a cigarette (62 μg/kg) is inhaled, but no data are available on the amount absorbed. Assuming an absorption of 50% and a consumption of 20 cigarettes/day, 375 ng/day would be absorbed (BGA, 1979). Based on data of the thallium concentration in urine of about 120 people, non-exposed individuals and workers with suspected industrial exposure to thallium, Apostoli et al. (1988) found no evidence of a difference between smokers and non-smokers, all about 40 years old.

Comparing a total of 128 men, no correlation was found between duration of employment at a cement plant (1 to 42 years) or age (16 to 62 years) and thallium concentrations in the urine (< 0.3 to 6.3 μg thallium/g creatinine) (Schaller et al., 1980). Therefore, it can be concluded that the uptake of low amounts of thallium does not cause accumulation in the body.

6.2.2.3 *Transplacental transfer of thallium*

Abortion was produced in the fourth month of pregnancy 8 days after ingestion of approximately 750 mg thallium sulfate. Starting 2 days after the ingestion, the mother was treated (haemodialysis, forced diuresis, Prussian Blue) for 92 h and survived. Before the start of the therapy, the blood of the mother and of the fetus contained 0.07 and 0.01 mg/litre and the urine 0.4 and 0.1 mg/litre, respectively. One day after the end of the treatment, the bones, liver and kidney of the fetus contained 0.9, 0.8 and 0.3 mg thallium/litre, respectively, and the blood of the mother 0.08 mg/litre (Graben et al., 1980). Additional evidence for the transplacental transfer of thallium is provided by studies demonstrating effects in infants exposed *in utero* (section 8.5.1).

6.3 Metabolic transformation

Data on the transformation and the equilibrium between the two oxidation states of thallium ions(I and III) in body fluids and tissues of mammals are not available. The two ions show a similar intracellular distribution (Sabbioni et al., 1980b).

6.4 Elimination and excretion

6.4.1 Animals

In a study on the accumulation and excretion of thallium in mussels and clams (section 6.1.1.1), the bivalves needed 7 and 30 days, respectively, to excrete all absorbed thallium. This is rapid in comparison to other heavy metals, such as cadmium, copper, lead and mercury, so that no significant amounts of thallium should enter the food web in this way (Zitko & Carson, 1975).

Within 25 days after parenteral administration of 10 mg thallium sulfate/kg body weight, rats eliminated 26% in the urine and 51% in the faeces (Lund, 1956a). Elimination via urine started within hours after oral application and persisted for up to 3 months. Faeces were not found to contain thallium until the fourth day, and thallium was still present after 1 month (Oehme, 1978). After injections of low doses of ^{204}thallium nitrate by different routes, the ratio of faecal to urinary elimination of rats increased with time from 2 to 5 (Lie et al., 1960). Gregus & Klaassen (1986) reported that faecal elimination is always greater than renal elimination. Biliary elimination is of minor importance (Schäfer & Forth, 1980; Gregus & Klaassen, 1986). Within 4 h after an intravenous injection, less than 0.3% of the injected thallium was eliminated in the bile of rats, but up to 8% into the gut (Sabbioni et al., 1984a). An even lesser degree of elimination occurred in tears, sweat, and milk (Oehme, 1978).

In contrast to absorption, secretion of thallium into the gut of rats (given as ^{201}thallium sulfate intravenously or directly into the individual ligated gastrointestinal segments) is highest in segments of the jejunum, followed by the ileum, colon and stomach (Sabbioni et al., 1984a). Similar results were obtained after intravenous administration to rats: *in situ*, the jejunum showed the highest excretory activity, followed by the ascending colon (Henning & Forth, 1982). The ileum and descending colon each excreted about half of the amount of the jejunum; excretion into the stomach was negligible. An increased dose (4 to 400 μg of thallium(I) sulfate/kg body weight) caused increased excretion into the jejunum and descending colon (Henning & Forth, 1982). Since thallium is also absorbed in the colon (section 6.1), only a proportion of the secreted thallium appears finally in the faeces.

Thallium ions are secreted against an electrochemical or concentration gradient by an active transport mechanism, as shown in experiments on the isolated mucosa of the descending rat colon (Schäfer et al., 1981; Schäfer & Forth, 1987). Thallium(I) ions use, at least in part, the same transport systems as potassium (Henning & Forth, 1977; Schäfer & Forth, 1987), and thallium secretion is reduced when the concentration of potassium is increased on the serosal side (Henning et al., 1982).

6.4.2 Humans

The normal daily total elimination in humans is estimated to be in the range of 1.64 μg thallium (urine: 1.2 μg; hair: 0.32 μg; faeces: 0.06 μg; skin and sweat: 0.06 μg) (US EPA, 1980). About 50% of total urinary elimination occurs within 9 to 11 days (Weinig & Schmidt, 1966).

In lethal cases of human poisoning, postmortem examination has always demonstrated high concentrations of thallium, especially in the contents of the colon (Table 21). Thallium levels in human saliva are up to 15 times higher than in the urine during the initial 2 weeks (Richelmi et al., 1980). Minor amounts are eliminated via hair and nails, both of which show the highest thallium concentrations of any tissue among human populations in uncontaminated areas (Table 19). Usually mother's milk is not an important route of elimination for heavy metals (Hapke, 1988). However, 2 weeks after a suicide attempt using thallium sulfate following the birth of her child (about 500 mg thallium), the milk of a mother contained 0.25 mg thallium/litre, while her blood only contained 0.07 mg/litre (Graben et al., 1980).

6.4.3 Methods to estimate daily intake of thallium

There are two ways to estimate daily intake of thallium, one based on total daily excretion and the other on the total amount of thallium in the body. In the former case, the total amount of thallium excreted daily under steady-state conditions (a model which may be reasonably applied to long-term exposure to low doses of thallium) should reflect accurately the daily intake of thallium. Using a mean urinary concentration of 0.4 μg/litre (which has been frequently reported in unexposed populations), a daily urinary excretion of 0.6 μg may be calculated assuming a urinary volume of 1.5 litre per day. Since we have assumed that renal excretion may account for about 70% of the total daily excretion of thallium, another 0.3 μg/day would be excreted by other routes, giving a value for total thallium daily intake of about 0.9 μg. A similar procedure leads to an estimated thallium daily intake of about 11 μg in chronically exposed populations (using a mean urinary concentration of 5 μg/litre).

The other method for estimating daily intake assumes that the following relationship exists between the total amount of thallium in the body (A_b), the daily intake of thallium (A_d) and the elimination rate constant (K):

$$A_d = KA_b$$

Since the total amount of thallium in the body has been estimated to be 100 μg per 75 kg body weight in an unexposed population (Weinig & Zink, 1967), a daily intake of 2.3 μg may be calculated, assuming an elimination rate constant of 0.023 day^{-1}.

6.5 Retention and turnover (biological half-life)

6.5.1 Animals

The biological half-life of thallium in experimental animals is 3 to 8 days. Accordingly, the elimination of 70 to 90% of the administered dose takes about 4 weeks (Oehme, 1978).

Using different routes of administration of thallium-204 in rats, Lie et al. (1960) found a biological half-life of 3.3 days during the first 21 days or until about 1% of the administered dose remained in the body. This body clearance was not affected by the route of administration and did not differ between various organs, except for the hair, which contained up to 60% of the body burden after 21 days. A biological half-life of 24 h was determined in pregnant mice and rats (Gibson et al., 1967).

Durbin et al. (1957) determined a biological half-life in rats of 5.2 days and calculated half-lives of 7 and 6 days for removal from the kidney and muscle, respectively. The half-lives in various organs (brain, spinal cord, sciatic nerve, kidney, liver and spleen) were lower in young rats than in adult rats and varied in different organs of young and adult rats. Ducket et al. (1983) found that half-life values in young rats ranged from 1.2 days for the sciatic nerve to 5.1 days for the liver (average for all tissues: 2.6 days) and in adult rats from 2.7 days for the brain to 6.0 days for the spleen (average: 3.8 days).

6.5.2 Humans

Several investigators have reported on the half-life of thallium in plasma and whole blood of humans acutely poisoned. Hologgitas et al. (1980) reported the half-life in the blood of one patient to be 1.9 days. Heath et al. (1983) reported a half-life in the blood of one patient of 1.9 days and a range of 21-24 h for the half-life in blood for three patients. Treatment for thallium toxicosis has been found to decrease the half-life of thallium in blood. In a review by de Groot & van Heijst (1988), the half-life in blood decreased as follows:

Treatment	*Thallium half-life in blood*
No treatment (n=2)	9.5; 15 days
Prussian Blue (PB) (n=5)	3.0 ± 0.7 days
PB + forced dialysis (FD) (n=7)	2.0 ± 0.3 days
PB + FD + haemoperfusion (n=3)	1.4 ± 0.3 days

In cases of poisoning, the half-life of thallium in blood is found to increase somewhat with time. Starting measurements at 42 days after toxic ingestion of thallium, Chandler et al. (1990) reported a blood half-life of 3.7 days in a patient treated with PB and intravenous potassium. The Gauss-Newton optimization model was used in this calculation. Wainwright et al. (1988) and Schwartz et al. (1988) presented data showing similar half-lives for thallium in urine and in serum, but no quantitative analysis was performed.

There has only been one study of the whole-body half-life of thallium in normal (i.e. unpoisoned) humans. In an investigation into the use of radiolabelled thallium for medical imaging, Atkins et al. (1977) administered thallium-200 to three volunteers. Using a whole-body counter, the biological half-life for thallium was found to be 9.8 days (range = 7.4-12.4 days). This determination is of much greater value than the determinations of plasma or whole blood half-lives for evaluating total excretion of thallium from the body.

6.6 Kinetics at the cellular level

The cellular uptake of thallium has been investigated in various systems. Due to the similarity in ionic radius of thallium(I) and potassium, thallium can substitute for potassium in a variety of potassium-dependent transport processes, as indicated by studies with microorganisms and frog skin (Norris et al., 1976; Zeiske & van Driessche, 1986). In rats and dogs, data indicate that "the mechanism involved in the active transport of potassium cannot differentiate between potassium and thallium" (Gehring & Hammond, 1967) (section 7.11).

The cytosol contains most of the intracellular thallium. In rats, autoradiography revealed the presence of thallium in the cytoplasm of nervous tissues during the first few days after injection (Ducket et al., 1983), a phenomenon also evident in kidney, liver and testis homogenates of rats treated with oral or intraperitoneal doses of 0.00004, 2, 20, 2000 or 3150 μg thallium(I) per rat (Sabbioni et al., 1980a,b) and in mussels (Nolan et al., 1984).

In a postmortem examination of a fatal case of thallium poisoning, 87% of the thallium was present in the cytosol (Davis et al., 1981) (for data on plants see section 4.1.2.3).

7. EFFECTS ON LABORATORY MAMMALS AND *IN VITRO* TEST SYSTEMS

The reported toxic effects of thallium are not always comparable, since no standardized procedure is used by the different authors and the duration of the experiments is not always stated. However, the effective dose (ED) of thallium at which minimal adverse effects (LD_{min}) occur, or at which 50% or all organisms are killed (LD_{50} and LD_{100}, respectively), are clearly correlated with the duration of the experiment. This correlation also exists for the period of time at which, for example, 50% of the organisms are killed.

The sections on single exposure (section 7.1.2), short-term exposure (section 7.2.2) and long-term exposure (section 7.3.2) cover the effects on various organs, except those on skin and eye (section 7.4) and the nervous system (section 7.8).

7.1 Single exposure

7.1.1 *Toxicity and symptoms*

The acute toxicity data for thallium compounds are listed in Table 23. They vary considerably with observation time, e.g., a 7-fold higher LD_{50} value (2000 mg/kg) for mice is obtained if the period of observation is reduced from 24 h to 1 h (Achenbach et al., 1980).

There exist only insignificant differences in the toxicity of various water-soluble thallium(I) salts to mice, rats, rabbits and dogs. In general, for most laboratory species and an observation period of 1 week, the LD_{50} or minimum effective dose (MED) values range between 20 and 60 mg/kg body weight for thallium(I) salts, independent of the application route, with the exception of guinea-pigs (5 to 15 mg/kg). The toxicity of water-soluble thallium compounds is similar for oral and parenteral routes of administration, indicating a high degree of gastrointestinal absorption (Table 23). The toxicity of water-insoluble thallium(III) oxide in rats and rabbits is 2 to 4.5 times higher following oral administration than following parenteral administration (Downs et al., 1960).

Acute toxicity is characterized by severe symptoms and/or death, which may be caused by single exposure or by multiple

Table 23. Toxicity of different thallium compounds in experimental animals after single exposure

Species	Thallium compound	Route of administration	Toxicity[a]	Period of observation	Dose (mg/kg body weight) Thallium compound	Dose (mg/kg body weight) Thallium ion	Reference[b]
Mouse	(I) carbonate	subcutaneous	LD_{min}	14 day	18	15.7	Sanotskii (1961)
Mouse	(I) chloride	intraperitoneal	LD_{50}	-[c]	24	20.5	Luckey & Venugopal (1977)
Mouse	(I) sulfate	oral	LD_{50}	1 h	2470	2000	Achenbach et al. (1980)
Mouse	(I) sulfate	oral	LD_{50}	2 h	1358	1100	
Mouse	(I) sulfate	oral	LD_{50}	4 h	988	800	
Mouse	(I) sulfate	oral	LD_{50}	12 h	432	350	
Mouse	(I) sulfate	oral	LD_{50}	24 h	370	300	
Mouse	(I) sulfate	oral	LD_{50}	36 h	235	190	
Mouse	(I) sulfate	oral	LD_{50}	-	30	24.3	IPS (1982)
Mouse	(I) sulfate	intraperitoneal	LD_{50}	10 days	47.5	38.5[d]	Stavinoha et al. (1959)
Mouse	(I) sulfate	intraperitoneal	LD_{10}	10 days	37	30	
Rat	(I) acetate	oral	LD_{min}	14 days	37.4	29	Downs et al. (1960)
Rat	(I) acetate	oral	LD_{50}	7 days	41.2	32	
Rat	(I) acetate	intraperitoneal	LD_{min}	14 days	25.8	20	
Rat	(I) acetate	intraperitoneal	LD_{50}	7 days	29.6	23	
Rat	(I) nitrate	intravenous	LD_{50}	-	16.3	12.5[e]	Gehring & Hammond (1967)
Rat	(I) nitrate	intravenous	LD_{50}	-	18.9	14.5[e]	Gehring & Hammond (1967)

Table 23 (contd).

Rat	(III) oxide	oral	LD_{min}	14 days	22.3	20	Downs et al. (1960)
Rat	(III) oxide	oral	LD_{50}	7 days	43.6	39	
Rat	(III) oxide	intraperitoneal	LD_{min}	14 days	103	92	
Rat	(III) oxide	intraperitoneal	LD_{50}	7 days	80.5	72	
Rat	(I) sulfate	oral	LD_{50}	-	10-25	8.1-20.2	IPS (1982)
Rat	(I) sulfate	dermal	LD_{50}	7 days	500	405	IPS (1982)
Rat	(I) sulfate	intraperitoneal	LD_{100}	2-6 days	33.3	27	Nachman & Hartley (1975)
Rat (*R. norvegicus*)	(I) sulfate	oral	LD_{50}	-	15	12.1	Wegler (1970)
Rat (*R. rattus*)	(I) sulfate	oral	LD_{50}	-	76	61.5	Wegler (1970)
Hamster	(I) malonate	oral	LD_{50}	> 7 days	50	40	Aoyama et al. (1988)
Hamster	(I) malonate	oral	LD_{100}	5-7 days	62.5	50	
Rabbit	(I) acetate	oral	LD_{min}	14 days	24.5	19	Downs et al. (1960)
Rabbit	(I) acetate	intravenous	LD_{min}	14 days	25.8	20	
Rabbit	(I) acetate	intraperitoneal	LD_{min}	14 days	16.8	13	
Rabbit	(III) oxide	oral	LD_{min}	14 days	33.5	30	Downs et al. (1960)
Rabbit	(III) oxide	intravenous	LD_{min}	14 days	43.6	39	
Rabbit	(III) oxide	intraperitoneal	LD_{min}	14 days	67	60	

Table 23 (contd).

Species	Thallium compound	Route of administration	Toxicity[a]	Period of observation	Dose (mg/kg body weight) Thallium compound	Thallium ion	Reference[b]
Guinea-pig	(I) acetate	subcutaneous	LD_{100}	5-9 days	19.3	15	Kaeser & Lambert (1962)
Guinea-pig	(I) acetate	oral	LD_{min}	14 days	15.5	12	Downs et al. (1960)
Guinea-pig	(I) acetate	intraperitoneal	LD_{min}	14 days	9.0	7	
Guinea-pig	(III) oxide	oral	LD_{min}	14 days	5.6	5	Downs et al. (1960)
Guinea-pig	(III) oxide	intraperitoneal	LD_{min}	14 days	33.5	30	
Dog	(I) acetate	oral	LD_{min}	14 days	25.8	20	Downs et al. (1960)
Dog	(III) oxide	oral	LD_{min}	14 days	33.5	30	Downs et al. (1960)

[a] LD_{min} = minimal dose at which lethality was observed; LD_{10} = dose at which 10% of the animals were killed; LD_{50} = dose at which 50% of the animals were killed; LD_{100} = dose at which all animals were killed

[b] Selected data; detailed data summarized in Negherbon (1959), Christensen et al., (1973), Zitko (1975a, b),Smith & Carson (1977), Venugopal & Luckey (1978), Schoer (1984), Nessler (1985a), Manzo & Sabbioni (1988), ATSDR (1992)

[c] No data given

[d] Graphical determination based on data of authors

[e] Rats fed a low or high potassium diet. An increase in dietary potassium had a small protective effect.

lower doses administered within 24 h. These symptoms are associated with disorders of the digestive (vomiting, diarrhoea) and nervous system, inflammation at body orifices, skin furuncles, tremor, loss of hair, a necrotizing renal papillitis, and death by respiratory failure (BGA, 1979; Hapke, 1984; Kazantzis, 1986; Bruère et al., 1990).

In rats oral administration of a toxic dose of thallium sulfate (10 mg thallium/kg) caused reduced food intake, diarrhoea, lethargy and ocular haemorrhage regardless of whether or not the animal survived the first 72 h (Sabbioni et al., 1982).

Summarizing 34 cases of canine thallotoxicosis, Zook & Gilmore (1967) emphasized the variability of the sequence and the severity of symptoms, partly due to the intoxication stage. Frequent symptoms were vomiting (82%), cutaneous alterations (71%), depression (62%), anorexia (53%), nervous disorders (47%), diarrhoea (44%), respiratory difficulty (44%), conjunctivitis (41%), dehydration (24%) and oesophageal paralysis (6%). The sequence in which the symptoms of intoxication occurred was generally as follows: first anorexia, vomiting and depression, then skin changes, dyspnoea and nervous disorders. Usually, rectal temperature was not elevated but was later often subnormal. After 3 to 7 days of illness, erythematous lesions occurred, which were most severe near mucocutaneous junctions and on the foot pads. The haematological findings were leukocytosis (neutrophilia, lymphopenia, eosinopenia) and haemoconcentration. Proteinuria and bilirubinuria were commonly observed. Autopsy and histopathological examination of 15 and 12 dogs, respectively, revealed increased heart weight (presumably caused by systemic hypertension), myocardial necrosis, congestion of the kidney with tubular nephrosis, pulmonary oedema, enlarged spleen, enlarged or oedematous lymph nodes, dilatation and areas of erosion of the oesophagus and necrosis of skeletal muscles. Some myelinated nerves showed focal distensions of their myelin sheaths, and lesions of the cerebrum and cerebellum were evident in dogs with neurological disorders (Zook & Gilmore, 1967). In addition, alopecia, anorexia, emesis and tenesmus (Coyle, 1980) and haematemesis (Waters et al., 1992) have been reported in cases of acute poisonings in dogs.

In cats similar signs of thallium poisoning have been found, e.g., skin alterations, apathy, lack of appetite, vomiting and signs of peripheral and central neuropathy (Zook et al., 1968; Fitzek & Henning, 1976). Haematological and histopathological findings were similar to those obtained with dogs. In older cats that died early, haemorrhagic gastroenteritis and hepatic or renal damage were evident (Zook et al., 1968).

Implantation of a pellet of pure thallium (3 to 5 mm diameter) into the motor cortex of a monkey, *Macaca mulatta*, resulted in death within 6 days (Chusid & Kopeloff, 1962).

7.1.2 Effects on various organs

Effects on the various organs have been summarized by Sabbioni & Manzo (1980) and ATSDR (1992). In nearly all affected organs direct cytotoxic effects as well as indirect effects due to damage of the nervous system have been found.

In acutely poisoned rats (single subcutaneous injection of 20 to 50 mg thallium acetate/kg), there was mild to moderate enteritis, including oedema of the submucosa and muscularis layers, and moderate to severe colitis (Herman & Bensch, 1967). Ultra-structural degenerative changes in the liver were frequently present, especially in the mitochondria. These were also indicated by increased numbers of autophagic lysosomes and lipid droplets (Herman & Bensch, 1967) and were evident 16 h after intraperitoneal injection of 50, 100 or 200 mg thallium(III) chloride/kg into rats (Woods & Fowler, 1986). It was concluded that "thallium-induced alteration of hepatic functional processes may arise from physical disruption of the membranal integrity of subcellular organelles with which those processes are functionally associated" (Woods & Fowler, 1986). In an extreme case of a poisoned dog, all the parenchymal hepatocytes were necrotic and the adrenals, thyroid, pituitary and pancreas had degenerated (Larson et al., 1939).

The kidney is considerably affected in poisoned animals (section 9.3.3). Since the concentration of thallium in the kidney is higher than in other organs, it would seem to be a specific target organ. Light microscopy of rat and mouse kidney tissue up to 48 h after administration of lethal doses of thallium (30 to 40 mg thallium sulfate/kg body weight) showed stromal oedema, necrosis of the loops of Henle and hydropic degeneration, as well as swelling and focal necrosis of the epithelium of proximal convoluted tubules (Danilewicz et al., 1979, 1980). Electron microscopy revealed degenerative changes in the epithelial cells of the glomeruli and tubules. The same changes in the kidneys of acutely poisoned rats have been described by Herman & Bensch (1967) (section 7.1.2).

Oral and intravenous administration of thallium sulfate (20 to 40 mg/kg) to the dog, cat, rabbit, goat and pigeon also caused direct effects on the respiratory apparatus, in addition to decreasing vasomotor reactivity (Rossi et al., 1981).

In *in vitro* studies, addition of 30 to 50 μg thallium sulfate reduced the beat frequency in isolated frog hearts (Buschke & Jacobsohn, 1922). Over a range of thallium concentrations, from those encountered after uptake from a contaminated environment (2 μg/litre) to those seen after suicide or homicide (and also higher levels of up to 204 mg/litre), the contractility of sheep interventricular cordis muscles exhibited three types of response, but they were not correlated to the thallium concentrations or period of incubation (Ziskoven et al., 1982). Using guinea-pig papillary muscles and low concentrations (0.02 to 2 mg thallium/litre), positive inotropic transients were followed by an inotropic decay (Ziskoven et al., 1982). In contrast, thallium produced concentration- and time-dependent positive inotropic effects in guinea-pig atrial preparations, but also inhibition of the sodium pump in ventricular slices (Ku et al., 1978). However, there is no discrepancy between these two effects, assuming that thallium inactivates the already fully activated pump and stimulates the inactivated pump (Ziskoven et al., 1982). The authors suggested that at low concentrations the effects of thallium are not associated with changes of membrane activity but with energy supply. Parameters of the slow inward current at the membrane level were not specifically altered by thallium (Wiemer et al., 1982). Another investigation involving guinea-pig papillary muscles and sheep Purkinje fibres indicated that the arrhythmogenic effects of thallium are restricted to the sinus node (Achenbach et al., 1982). High concentrations (200 mg/litre) depolarize the muscle fibre membrane and lead to irreversible damage (Mullins & Moore, 1960). Within the muscle, thallium seems to compete for the adsorption sites normally occupied by potassium, being adsorbed on to myosin and thus being localized primarily in the A band (Ling, 1977).

7.2 Short-term exposure

7.2.1 *Toxicity and symptoms*

In mice the daily supplementation of food with 400 μg thallium acetate induced alopecia after about 14 days, followed by increasing apathy and death within 16 to 18 days after the beginning of the treatment (Buschke, 1900).

Daily intraperitoneal injection of thallium(I) acetate (5 mg/kg) for 7 days in rats caused anorexia, reduced growth, irritability and tenderness during handling, lethargy, diarrhoea, dragging of the hind limbs, fits of abnormal rotation of head and neck and curving of the body. About 15% of the rats died (Hasan et al., 1977c).

7.2.2 Effects on various organs

In subacutely poisoned rats (2 to 3 injections of 10 to 15 mg thallium acetate/kg at intervals of 1 week), only slight colitis and enlargements of mitochondrial granules in the liver occurred (Herman & Bensch, 1967). Electron microscopy of the kidney revealed similar changes to those observed in acutely poisoned animals (section 7.1).

7.3 Long-term exposure: chronic toxicity

7.3.1 Toxicity and symptoms

The chronic toxicity data on thallium compounds are listed in Table 24.

In rabbits daily administration of thallium(I) sulfate or carbonate (0.25 mg/kg) for 6 months caused disturbed behaviour, aggressiveness, diarrhoea, and loss of hair (Tikhova, 1967). In rats daily ingestion of 7.5 mg thallium was lethal (Bowen, 1966).

In a study of weanling albino rats (50-80 g) fed *ad libitum* for 1 month on a diet containing 2, 10, 50, 100, 500 or 5000 mg thallium(I) acetate per kg diet (5 rats/group), dietary levels of 2 and 10 mg/kg diet caused no effects on growth and survival within the feeding period, whereas the other concentrations resulted in mortalities of 60 to 100% within 10 days. When rats were fed for 15 weeks with 5, 15, 30 or 50 mg thallium acetate/kg diet (5 males and 5 females/group), the two lowest doses did not affect growth of males or females. The 15 and 30 mg/kg doses caused hair loss starting after 2 weeks, and after 15 weeks the rats were almost free of hair. Within the fourth and eighth week and after intakes of 30 mg/kg diet, 80% of males and 60% of females died. At 50 mg/kg all males died within 2 weeks and all females within 8 weeks. No specific pathological alterations were found in any organ. A dose of 30 mg/kg diet resulted in moderate growth depression in males but not in females, while in both sexes increased mortality was observed. For thallium(III) oxide the effective concentrations were similar, and males also reacted more

Table 24. Toxicity of different thallium compounds in experimental animals after several administrations

Species	Thallium compound	Route of administration	Toxicity[a]	Period of observation	Dose (mg/kg body weight)		Reference
					Thallium compound	Thallium ion	
Mouse	(I) chloride	intraperitoneal	LD_{50}	30 days	1.2	0.1	Bienvenu et al. (1963)[b]
Mouse	(III) chloride	intraperitoneal	LD_{10}	30 days	6.0	4.0	Hart & Adamson (1971)[b];
			LD_{50}	30 days	6.9	4.5	Adamson et al. (1975)
Mouse	(I) nitrate	intraperitoneal	LD_{50}	14 days	37.5	28.6	Williams et al. (1982)[b]
Rat	(III) chloride	intraperitoneal	LD_{10}	30 days	4.85	3.2	Hart & Adamson (1971)[b];
			LD_{50}	30 days	5.66	3.7	Adamson et al. (1975)
Rat	(I) sulfate	oral	LD_{20}	15 days	1.25	1.0	Tikhova (1964)
			LD_{min}	15 days	0.6		
Rabbit	(I) carbonate	oral	LD_{min}	180 days	0.25	0.2	Tikhova (1967)[c]
Rabbit	(I) carbonate	subcutaneous	LD_{min}	180 days	0.25	0.2	
Rabbit	(I) sulfate	oral	LD_{min}	180 days	0.25	0.2	Tikhova (1967)[c]
Rabbit	(I) sulfate	subcutaneous	LD_{min}	180 days	0.25	0.2	

[a] LD_{min} = minimal dose at which lethality was observed; LD_{10} = dose at which 10% of the animals were killed; LD_{50} = dose at which 50% of the animals were killed; LD_{100} = dose at which all animals were killed

[b] Daily injection for 10 consecutive days

[c] Value in winter was lower than in summer

sensitively than females. No specific pathological alterations were found in any organ except the skin, where atrophy of hair follicles and sebaceous glands were seen at both higher dose levels. The exact concentration of thallium ingested by the rats could not be determined but was estimated to be in the range of 1 to 3 mg thallium acetate/kg body weight per day for the diet containing 15 mg thallium acetate/kg food (Downs et al., 1960).

US EPA (1986) conducted a 90-day study in which male and female Sprague-Dawley rats (20 of each sex per group) were administered aqueous thallium sulfate by gavage at doses of 0.01, 0.05 or 0.25 mg/kg body weight per day. Both untreated controls and vehicle (water) treated controls were included. Clinical observations were recorded daily and neurotoxicological examinations were performed 3 times per week on selected animals. Haematological and clinical chemistry parameters were measured on days 0, 30 and 90, and ophthalmological examinations were performed on days 0 and 90. Upon necropsy, selected organs were weighed. No significant differences were seen in any group for body or organ weights. Several changes were reported for blood chemistry parameters. Statistics were reported for male and female rats at 30 and 90 days for each dose group compared with both untreated controls and vehicle-treated controls. In both males and females, small increases were seen for serum glutamic-oxaloacetic transaminase, lactic acid dehydrogenase and sodium levels, with statistical significance at many points. At the lowest dose, statistically significant changes were seen in male rats for all three of these parameters, but only when compared with untreated controls. Higher doses resulted in statistical significance when compared with vehicle controls. Similar patterns were seen in female rats.

In addition, there was a dose-related increase in the incidence of alopecia, lacrimation and exophthalmos. No treatment-related changes were seen in the eye. Gross necropsy revealed only alopecia; this occurred in a dose-related manner for females and was apparent at the lowest dose level. In males, alopecia was also apparent in all dose groups but was not strictly dose-related.

In a study with female Sprague-Dawley rats (180-200 g), thallium sulfate was given via the drinking-water (10 mg thallium/litre) over a period of 40 weeks and with a total intake of about 80 mg thallium/rat (Manzo et al., 1983). First symptoms, i.e. poor hair lustre, periorbital redness and irritability, were observed on days 20 to 25. Hair loss first appeared after 32 days in some rats, and in several rats was almost complete by the end of the 36-week period of treatment. However, in other rats no hair loss occurred. Starting from day 40 the number of rats showing mild or severe cutaneous disorders increased strongly. After 40 days and a total ingestion of about 10 mg thallium/rat, lethality amounted to 15%, and the surviving rats showed no electrophysiological abnormalities. After 240 to 280 days and the ingestion of about 70 to 80 mg/rat only an additional 6% of the rats died; about two-thirds of the rats showed electrophysiological effects and reduced motor and sensory action potentials.

7.3.2 Effects on various organs

In chronically poisoned rats (initial injection of 10 to 20 mg/kg, followed by weekly subcutaneous injections of 5 mg/kg for 4 to 26 weeks), only slight colitis sometimes occurred (Herman & Bensch, 1967), but degeneration in the liver was similar to the severe effects in acutely poisoned rats (section 7.1.2) and liver enzymes were also affected (Bulavintseva & Bulavintsev, 1982). In the stomach of rats the production of hydrochloric acid was reduced (Buschke, 1929). Kidney weight was increased (Downs et al., 1960) and ultrastructure affected in the same way as in acutely poisoned animals (Herman & Bensch, 1967) (section 7.1.1). The greater accumulation of thallium in the renal medulla than in the renal cortex of chronically poisoned rats indicates a firmer binding of thallium, which might impede thallium elimination (André et al., 1960; Lameijer & van Zwieten, 1977a).

Intratracheal administration of 0.5 or 5 mg thallium(I) salts (iodide, bromide and chloride or mixtures of them) caused dose- and time-dependent changes in the lungs of rats, iodide being the most toxic salt (Spiridonova et al., 1978).

In chronically poisoned guinea-pigs, the adrenaline and lipoid contents of the adrenal glands were considerably, sometimes totally, reduced. Examination of chronically poisoned rats revealed a reduction in the size of the thyroid gland, follicular atrophy and some pycnotic nuclei (Buschke & Peiser, 1922a,b; Buschke, 1929).

7.4 Skin and eye irritation

7.4.1 Skin and hair

Thallium intoxications in dogs caused striking effects in all layers of the skin. The dermal changes were characterized by oedema and disruption of collagen bundles. In erythematous patches massive parakeratosis (much more pronounced than in other dermatoses) and occasionally a granular layer were found. In the follicles, of which only 60% contained hairs, the external root sheath was hyperplastic, showing an excess of parakeratotic horny material. Follicular plugging was prominent (Schwartzman & Kirschbaum, 1962). The progress of the erythematous lesions towards scaling and crusting included varying degrees of necrolysis and, characteristically for thallium intoxications, spongiform abscesses, the latter occurring also in the hair follicles (Schwartzman & Kirschbaum, 1962) where thallium binds strongly to melanin (Tjälve et al., 1982).

The mechanism of depilation is unclear. According to Truhaut (1960), this major sign of thallium intoxication, alopecia, is caused by its antimitotic activity, since hair follicles (and testes) are normally characterized by marked mitotic activity. This adverse effect is prevented by glutathione and cysteine. Counting the mitotic rate per hair follicle after a subcutaneous injection of 30 mg thallium(I) sulfate into young rats at 4 and 7 days of age, Cavanagh & Gregson (1978) observed an initial decline in mitotic rate followed by cell deaths in the matrix zone.

On the basis of results from early experiments on mice, rats, guinea-pigs, cats and rabbits performed by Buschke (1900, 1903, 1911), Buschke & Peiser (1922b,c, 1926) and Buschke et al. (1928), the authors put forward the hypothesis that the depilatory effect is mediated by the activation of sympathetic nerves, since light microscopy never revealed a direct effect on the hair follicles and since sensory hairs, innervated by cerebrospinal nerves, were never affected, in contrast to the smooth muscles of other hairs which are innervated by the sympathetic system.

In chronically poisoned rats (initial injection of 10 to 20 mg thallium acetate/kg followed by weekly subcutaneous injections of 5 mg/kg for 4 to 26 weeks) (Herman & Bensch, 1967), black speckling of periorbital hairs occurred occasionally. In other chronically poisoned rats (section 7.3) showing complete depilation, light microscopy revealed effects similar to those after acute intoxication, i.e. a keratinized epidermis, decrease in the size of sebaceous glands and reduction of the number of hair shafts and follicles, the latter becoming atrophic or replaced by scars, collagen or fat (Downs et al., 1960).

7.4.2 Eye

Autoradiography of adult mice demonstrated a relative accumulation of thallium in the lens of the eye (André et al., 1960). A biochemical investigation showed that in addition to binding to melanin in the hair follicles, thallium also binds strongly to melanin in the mouse eye. This might result in iritis and retrobulbar neuritis, which are regularly observed symptoms of thallium intoxication (Tjälve et al., 1982).

7.5 Reproductive toxicity, embryotoxicity and teratogenicity

Thallium affects reproduction in various ways. Buschke & Peiser (1922a) found a total reduction of sexual activity in intoxicated rats and mice, which did not develop after administration to young animals (Buschke & Peiser, 1922b,c) (section 7.3). This effect could be antagonized by the administration of an ovarial hormone and hypophyseal tissue (Buschke, 1929). Smith & Carson (1977) also emphasized that sexual activity is usually lessened in chronically poisoned animals.

7.5.1 Gonadotoxic effects

Several early, contradictory reports dealt with the question of whether or not thallium affects ovarian function or the estrous cycle (Smith & Carson, 1977). Buschke et al. (1927a) and Buschke & Berman (1927) described strong inhibition of the estrous cycle in mice.

More attention has been directed to the effects on the male reproductive system. Cytotoxic and perhaps mutagenic effects (section 7.6) can affect the offspring.

In several laboratory species, acute or repeated treatment with thallium(I) salts resulted in similar or even higher concentrations of thallium being found in the testes compared to other organs (section 6.2), pointing perhaps to a special susceptibility (Gehring & Hammond, 1967; Krassowski et al., 1977, 1980; Sabbioni et al., 1980a,b; Talas & Wellhöner, 1983; Aoyama, 1989). However, an effect on offspring has not been investigated in detail. Landauer (1930) demonstrated effects on chickens, but only two cocks were included. A second investigation, a dominant lethal test using rats, is discussed in section 7.6.

Several authors have reported histological findings. In acutely poisoned rats the epithelial cells of the seminal vesicle contained numerous autophagic vacuoles (Herman & Bensch, 1967), and in acutely intoxicated rabbits and dogs spermatogenesis was inhibited (Larson et al., 1939; Truhaut, 1960; Zook & Gilmore, 1967). Chronic poisoning caused total atrophy of testes in some rats (Buschke & Peiser, 1922a,b,c) or considerably disturbed spermatogenesis, which, however, returned to normal after termination of thallium administration (Buschke, 1929). A 6-month administration of thallium carbonate in the drinking-water of mice (0.001 mg/litre or 0.01 mg/litre) caused a decrease in sperm fertility at the high dose level and reduced sperm motility at the lower dose level (Wei, 1987). Increased desquamation of spermatogenic epithelium in rats was observed after oral administration of 0.001 mg thallium carbonate/kg (Shabalina et al., 1980). In rats treated with 10 mg thallium/litre in their drinking-water for 2 months, but not in those treated for only for 1 month, there was rearrangement of the germinative epithelium, premature release of germinal cells into the seminiferous tubules, low epididymal sperm motility, the appearance of immature elements in the semen and a high susceptibility in the Sertoli cells (Formigli et al., 1985, 1986; Grégotti et al., 1985), the latter also reacting very sensitively in *in vitro* studies (Grégotti et al., 1993). However, thallium treatment did not affect the absolute or relative weight of the testes or the plasma levels of testosterone. The mechanisms underlying the findings for thallium may involve its antimitotic effect or its effect on energy metabolism (Formigli et al., 1985, 1986). The use of thallium-201 (1-10 μCi), which is also utilized for diagnosis in medicine, for testicular imaging in mice caused loss of testicular weight and reduction in the number of sperm heads. These effects were less severe when thallium-204 (1-10 μCi) was used. This must have been due to the different radioisotopes (probably the low-energy Auger electrons of thallium-201), rather than to the physicochemical properties of thallium that determine uptake (Rao et al., 1983).

7.5.2 Embryotoxicity and teratogenicity

7.5.2.1 Chickens

Injections of thallium (1 or 2 mg per egg) into the yolk sac of 4-day-old chick embryos strongly affected growth and survival. Injections of thallium sulfate (0.7 mg/egg) into the chorioallantoic membrane caused similar effects (Karnofsky et al., 1950; Ridgway & Karnofsky, 1952).

More striking were the effects on the development of bones of the embryonic chick: achondroplasia, leg-bone curvature, parrot beak deformity and microcephaly (Karnofsky et al., 1950; Karnofsky & Lacon, 1964). The achondroplasia did not occur after injections of 0.2 mg thallium sulfate/egg into the yolk sac, but with the larger doses increasing percentages of the embryos showed shorter bones than normal, and the tibia and femora were strikingly curved. Thallium sulfate and nitrate showed a similar ability to produce achondroplasia (Karnofsky et al., 1950).

Achondroplasia was only produced if thallium treatment was performed during a critical, sensitive period (Hall, 1977, 1985). This period began on day 5 of incubation and ended after 205 to 207 h of incubation, coinciding with a 66% decrease in growth rate of the embryos. During this period thallium bound rapidly to skeletal tissues. Light and electron microscopy showed the primary site of action to be the cartilage, maturing hypertrophic chondrocytes being most affected (Ford et al., 1968; Hall, 1972; Skrovina et al., 1973). An increase of glucosamine and a decrease of the mucopolysaccharides without a simultaneous loss of collagen indicated an interference with cartilage metabolism (Ford et al., 1968). According to Hall (1972), thallium treatment caused an "abnormal distribution of the acid mucopolysaccharides in otherwise normal cartilage matrix and the formation of necrotic areas within the maturing hypertrophic chondrocytes". The reduced secretion of the acid mucopolysaccharide into the intercellular matrix did not affect ossification.

7.5.2.2 *Mammals*

In mammals, the results are inconsistent and there seems to be great variation between species and strains (Claussen et al., 1981). Intraperitoneal injection in rats on days 8 to 10 (2.5 mg thallium sulfate/kg) or on days 12 to 14 (2.5 or 10.0 mg/kg) of gestation had no effect on resorption rates but significantly reduced the body weight of fetuses examined on day 21 of gestation. Hydronephrosis and non-ossification of vertebral bodies were observed. The effects on fetal body weight and vertebrae were also induced by a low potassium diet and were not increased by an additional dose of thallium (Gibson & Becker, 1970). According to Barlow & Sullivan (1982), the increase in hydronephrosis may not be related to thallium. Adverse effects on the cartilage of long bones in 6- and 9-day-old rats treated intraperitoneally with thallium sulfate were described by Nogami & Terashima (1973). These results might explain the reduced growth of the suckling rats of thallium-intoxicated dams (Ehrhardt, 1927). Ossification of rat skulls was also reduced (Buschke et al., 1927b).

In a preliminary study on mice, oral administration of thallium sulfate (0.3, 1 and 8 mg thallium/kg body weight) to 3-6 pregnant SWS mice (gestation day 9) caused achondroplasia in 12.5, 14 and 50% of the offspring, respectively (Achenbach et al., 1979a,b). Earlier administration of thallium induced miscarriage, while older fetuses were only slightly affected. In other substrains of the SWS mice, administration of 10 to 20 mg/kg induced no teratological effects (Claussen et al., 1981). Under similar conditions, but using 29 NMRI mice and oral doses of 0.8 and 8 mg thallium/kg body weight, the weights of fetuses and placentas were not affected by the treatment with thallium sulfate, but a slight, statistically non-significant increase in the intra-uterine mortality rate was found in embryos examined 18 days after fertilization. A significantly increased rate of double placentas and of effusions of blood in the thigh of the fetuses was observed at 8 mg/kg (Török & Schmahl, 1982). Oral administration of 3 or 6 mg thallium chloride or acetate per kg per day to NMRI mice (about 30 per group; day 6 to 15 of gestation) had no observable effects on skeleton or organs at day 18 of gestation (Claussen et al., 1981). However, post-implantation losses were increased after administration of 6 mg thallium chloride/kg, but only young embryonic stages were affected. Administration of 6 mg thallium acetate/kg caused a reduction in the weight of embryos. In a parallel study with rats, 6 or 4.5 mg thallium chloride or acetate/kg was lethal to the dams (gestation day 6 to 15), while 3 mg/kg induced no embryotoxic effects but increased malformations of ribs and vertebrae (Claussen et al., 1981).

Studies of mouse embryos cultured *in vitro* showed that low levels of thallium (0.2 mg/litre) affected the pre-implantation stage (summarized by Formigli et al., 1985).

In vitro cultivation of mouse limb buds (day 11 of gestation) in medium containing 10 mg/litre for a period of 3 days, followed by cultivation in normal medium for another 3 days, caused impaired development of the hand skeleton (Barrach & Neubert, 1985). In a similar study, the extent of this effect was also increased by decreasing the potassium concentrations of the medium, and vice versa (Neubert & Bluth, 1985). The authors suggested that an interference of thallium with mammalian embryonic development is possible if its concentration in embryonic tissues reaches levels above 2-3 mg/kg.

When rat embryos (10.5 days after fertilization) were cultured for 48 h in media containing thallium sulfate (3, 10, 30 or 100 mg thallium/litre), embryotoxic effects were evident at all concentrations. Increasingly deleterious effects occurred on growth and on mesoderm and entoderm at concentrations from 10 to 100 mg/litre. Whereas macroscopically no or only minor abnormalities could be detected, histological examination of the embryos revealed that cell necroses in the brain developed even at the lowest concentration. At higher concentrations these effects increased, and necrosis was complete at 30 mg/litre. The authors doubted whether a suggested embryotoxic concentration of 1 mg/litre can be achieved without severe maternal poisoning (Anschütz et al., 1981).

7.5.2.3 *Delayed effects on development of offspring*

Many offspring of chronically poisoned rats showing total alopecia died during the first two days after birth. In the other pups, hair development was severely impaired (section 7.4), and these pups died within 4 to 5 weeks after birth. In a second litter all offspring showed alopecia, but none of them died so early (Buschke, 1911).

In studies by Claussen et al. (1981), oral administration of 6 mg thallium chloride/kg to NMRI mice dams (day 6 to 15 of gestation) reduced the initial increase in weight of the offspring and slightly increased the mortality rate during the first 3 weeks. In a parallel study with rats, oral administration of 3 mg thallium chloride or acetate per kg caused similar effects.

In rats, exposed prenatally by treating the dams with 0.01 g thallium sulfate/litre in their drinking-water, gestation and weight of offspring were not affected, but hair development was retarded during the first 60 days of life. At an age of 30 or 60 days, the hypotensive and hypertensive cardiovascular responses to acetylcholine or to isoprenaline and noradrenaline, respectively, were lower than the responses of control animals (Matera et al., 1986; Rossi et al., 1988). Comparisons with postnatally exposed pups showed that the reactivity of the developing vascular autonomic nervous system was also lowered (Rossi et al., 1988).

Oral doses of 0.1, 0.5 or 1.0 mg thallium/kg body weight (thallium sulfate) given to rats on days 6, 7, 8 and 9 of gestation impaired the learning ability of their adult offspring in operant behaviour tests. Postnatal administration of thallium (same doses) had no such effect (Bornhausen & Hagen, 1984).

7.6 Mutagenicity and related end-points

Only two microbiological investigations have been performed; these indicated no mutagenic effects (Claussen et al., 1981). Both used the Ames test. Dehnen, whose unpublished data were described by Claussen et al. (1981), investigated the effects of thallium(I) acetate and used 3.1 μg-29.2 mg thallium/plate and the *Salmonella typhimurium* strains TA98, 1535, 1537 and 1538. Similarly, the use of Ames tests with thallium chloride and acetate (doses not given) and *S. typhimurium* strains TA98, 100, 1535, 1537 and 1538, with and without metabolic activation, indicated no mutagenic effects (Claussen et al., 1981). According to these authors, no sister-chromatid exchange was found in bone marrow cells after oral administration of thallium chloride to Chinese hamsters (5 or 10 mg thallium/kg body weight; twice after 24 h; 8 animals per dose).

However, thallium(I) carbonate induced sister-chromatid exchange and chromosomal aberrations in one cell line and hypoxanthine-guanine phosphoribosyl-transferase gene mutation in another cell line (Zhang, 1988). In studies with embryonic fibroblasts of various mouse and rat strains, the same thallium(I) compound (0.5 to 46.9 mg/litre) caused a significant increase in the single-stranded DNA fraction after incubation of rat fibroblasts and cells of one mouse strain ($C_{57}Bl/6$), whereas cells of the CBA mouse strain were resistant to the same concentrations. In a test of survival and mutability of vaccinia virus in both mouse cell lines, the CBA cells showed increased survival of the virus, suggesting greater efficiency of the repair systems (Zasukhina et al., 1980, 1983).

In a dominant lethal test, white rat males received daily oral doses of 5, 50 and 500 ng thallium carbonate/kg body weight over 8 months. Thereafter they were mated with untreated females. On gestation day 20, 18 dams were killed. At the two higher doses there was a tendency towards an increase in embryonic mortality, whereas at the lowest dose the number of resorptions and post-implantation deaths were increased (Zasukhina et al., 1983). Using these data in a T-test, the differences between the means for the total embryonic death of treated and untreated mice were statistically significantly different ($p < 0.05$).

A mutagenic effect on sperm cells of rats was reported after oral administration of 0.0001 mg thallium carbonate/kg (Shabalina et al., 1980). However, the report lacked detail concerning the experimental set-up and the Task Group considered that it could not be used for evaluation of the health effects of thallium.

7.7 Carcinogenicity

No standardized carcinogenicity studies have so far been performed (ATSDR, 1992). Owing possibly to their cytotoxic effects, thallium salts may have a local antineoplastic effect in mice and rats (Hart et al., 1971; Hart & Adamson, 1971; Adamson et al., 1975).

7.8 Neurotoxicity

7.8.1 Central nervous system

7.8.1.1 Histology and ultrastructure

In the brains of subacutely poisoned rats, Herman & Bensch (1967) (section 7.1.2) occasionally found foci of perivascular cuffing, while the mesencephalon of two of the four rats contained an extensive region of acute necrosis. In addition, numerous lipofuscin bodies were sometimes present in neuron cytoplasm. This was also found to occur following chronic poisoning; electron microscopy showed effects on mitochondria (Herman & Bensch, 1967).

The brain of acutely poisoned guinea-pigs (subcutaneous injection of 15 or 18 mg thallium sulfate/kg) showed slight microscopic alterations, e.g., swelling of cells and vacuolization in the perikaryon (Tackmann & Lehmann, 1971). In the right parietal cortex of rats, microglia cells and alpha astrocytes were affected 24 h after an intraperitoneal injection of 40 mg/kg (Reyners et al., 1981). At 24 h after intraperitoneal injection of 32 mg/kg into newborn rats, the encephalon showed oedema and congestion; even after an additional 50 days, there was focal destruction of neurons and irregular fibrosis of the capillary vessels (Barroso-Moguel et al., 1990).

Ataxia and tremors are known to be associated with cerebellar lesions and both neurological disorders occur in cases of thallium intoxication. Ultrastructural alterations of the cerebellum, especially of the mitochondria, were evident in rats after poisoning by daily intraperitoneal injections of thallium acetate (5 mg thallium/kg for 7 days) (Hasan et al., 1978a). Effects in other brain regions indicate that the effects of thallium on the activity of endocrine glands may be mediated via changes in hypothalamic control (Hasan et al., 1977b). Using identical conditions, Hasan et al. (1977a) observed an apparent increase in the number of oligodendrocytes and suggested a correlation with thallium-induced neuronal chromatolysis described by other authors, since the usual functions of oligodendrocytes are the formation of myelin and the nutrition of neurons.

7.8.1.2 *Electrophysiological and biochemical investigations*

In addition to the direct effects of thallium on the cardiovascular and respiratory apparatus of the dog, cat, rabbit, goat and pigeon, indirect effects involving higher vasomotor and respiratory centres (but mostly dependent on a decrease in vasomotor reactivity) were found following acute poisoning (section 7.1.2) (Rossi et al., 1981).

Extra- and intracellular recording of central neuronal activity in hippocampal slice preparations from guinea-pigs and rats showed that thallium (20 to 40 mg/litre) seems to have a predominantly postsynaptic effect in hippocampal slice preparations, perhaps by exerting an unspecific influence on the intracellular metabolic mechanisms of the CA1 pyramidal cells (Lohmann et al., 1989). Daily intraperitoneal injections of thallium(I) acetate (4 mg/kg for 7 days) indicated that the electrophysiological parameters of noradrenergic transmission in rat cerebellum were reduced (Marwaha et al., 1980).

The association of the corpus striatum with the pathogenesis of the abnormal movements that have been reported after thallium intoxication is shown by an increased firing rate of the caudate neurons of rats, 3 to 4 h after intravenous injection of 10 mg thallium sulfate/kg (Hasan et al., 1977c). After daily intraperitoneal injections of 5 mg thallium(I) acetate/kg for 7 days (section 7.8.1), the protein content of the corpus striatum was significantly increased and the respective breakdown enzymes were depleted. However, the latter did not occur in the cerebrum (Hasan et al., 1977b,c).

Certain motor dysfunctions are known to be associated with a decrease in the level of brain dopamine, an aspect supported by the data of Hasan et al. (1978b). Convulsive disorders may also be related to a brain deficiency of gamma-aminobutyric acid (GABA)-ergic mechanisms. Data obtained by Hasan et al. (1977d), but not those by Nisticò et al. (1984), support this interpretation. Neurotoxicity could also result from changes in the concentrations of amino acids and other neurotransmitters (Ali et al., 1990) or an acceleration of monoamine catabolism (Osorio-Rico et al., 1994).

Further important mechanistic aspects were increases found in the lipid peroxidation rates and in the activity of the lysosomal enzyme β-galactosidase, especially in the cerebellum, brainstem and cortex, after daily intraperitoneal injection of 8 mg thallium(I) acetate/kg for 6 days in rats (Brown et al., 1985). Also a lower dose of 4 mg/kg selectively altered patterns of behaviour (section 7.8.1.3). An increased deposition of lipofuscin-like pigment granules in the cerebellar neurons and an increase in lipid peroxidation rates in the cerebrum and brainstem of rats (which was even exceeded by that in the cerebellum) were described in a previous study (Hasan & Ali, 1981). This seems to be an important mechanism of toxicity (section 7.9).

7.8.1.3 Behavioural toxicology

Intraperitoneal injections of 10 or 20 mg thallium sulfate/kg produced only a slight conditioned flavour aversion in rats, perhaps due to the delayed onset of symptoms (Nachman & Hartley, 1975; Peele et al., 1986). Oral administration induced a dose-dependent aversion to saccharin at all but the lowest dose of 2.5 mg/kg (Peele at al., 1986, 1987). This difference might be explained by irritation of the gastrointestinal tract after oral uptake of thallium.

In a detailed investigation of the effects of daily intraperitoneal injections of 4 or 8 mg thallium(I) acetate/kg for 6 days on several behavioural patterns of rats, changes of behaviour were intensified with increased dose. Some of the changes correlated with biochemical effects (section 7.8.1.2) (indicating cellular damage) in certain regions of the brain (Brown et al., 1985).

In pest control campaigns against wild rodents, the dying animals left their hiding places and came to the surface, presumably due to extreme thirst and breathing disturbances (Larson et al., 1939).

7.8.2 Peripheral nervous system

In the final phase of lethal intoxication and *in vitro*, a parasympathetic stimulation seems to occur. Because thallium diminishes the effects of adrenaline on isolated hearts or intestine, even after parasympathetic blockage, it is thought that it may destroy adrenaline (Truhaut, 1960).

In the initial phase of intoxication, sympathetic nerves are stimulated (Buschke & Peiser, 1922b). Investigating the effects of 0.2 to 204 mg thallium(I) and thallium(III) ions/litre on the ATPases of the amine-storing granules from bovine adrenal medulla and splenic nerves, a specific inhibition by thallium(III), but not by thallium(I), was observed at concentrations which might occur in the tissues after intoxication with thallium(I). The authors suggested that thallium(I) is oxidized to thallium(III) in the organism. Because the ATPase of the nerve granula which store noradrenaline was nearly ten-fold more sensitive than that of the adrenal medulla, which stores mainly adrenaline, this might explain the strongly increased elimination of noradrenaline (Burger & Starke, 1969).

7.8.2.1 *Histology and ultrastructure*

In the peripheral and optic nerves of acutely, subacutely and chronically poisoned rats (Herman & Bensch, 1967) (section 7.8.1), no consistent or even slight changes were revealed by light or electron microscopy. However, partial atrophy of the optic nerve was found by Buschke et al. (1928). In addition, acute poisoning of a dog (Greving & Gagel, 1928) and guinea-pigs (subcutaneous injection of 15 or 18 mg thallium sulfate/kg) (Tackmann & Lehmann, 1971) caused alterations of the axons and myelin sheaths which were evident under the light microscope. In a 36-week study (Manzo et al., 1983) (section 7.3), in which rats were given thallium sulfate in their drinking-water (10 mg thallium/litre), about 50% of the animals developed Wallerian degeneration (myelin debris and vacuolization); lamination of the myelin sheath of the sciatic nerve fibres was confined to some large and medium-sized fibres. Degenerative lesions found in the white matter of the spinal cord of poisoned rabbits may account for the paralysis of the hind legs (Truhaut, 1960).

In a number of *in vitro* studies, thallium affected nerves, e.g., cell outgrowth was inhibited (Sharma & Obersteiner, 1981; Windebank, 1986) or the myelin sheath disintegrated (Peterson & Murray, 1965).

7.8.2.2 Electrophysiological and biochemical investigations

Following a subcutaneous injection of 15 mg thallium sulfate/kg to guinea-pigs, within days the larger and faster conducting nerve fibres degenerated before the slower fibres and became inexcitable (Kaeser & Lambert, 1962; Tackmann & Lehmann, 1971). In a subchronic study (section 7.3) on rats, which received thallium sulfate via their drinking-water (10 mg thallium/litre), the motor and sensory action potential amplitudes were unaffected after 40 days of poisoning but decreased after 240 days (Manzo et al., 1983). Then the motor action potential latency was increased, and no fibrillation activity was observed in the tibialis anterior muscle.

Neuromuscular transmission in thallium-treated rats and mice has been investigated in detail using phrenic nerve-diaphragm preparations (Wiegand et al., 1983, 1984a, 1986). The relationship between thallium concentration (z, in mM) and duration of paralysis (y, in minutes) is approximated by the equation $y = 4.6 \times e^{8.4z}$ (Csicsaky & Wiegand, 1981). It has been suggested that thallium interferes presynaptically with spontaneous transmitter release by antagonizing these calcium-dependent processes, rather than by interfering with the presynaptic influx of calcium ions (Wiegand et al., 1983, 1984a,b; Wiegand, 1988). Additional data indicate that thallium, like other heavy metals (Cooper et al., 1984), irreversibly blocks phasic transmitter release, while spontaneous transmitter release is reversibly enhanced (Wiegand et al., 1986; Csicsaky et al., 1988). The sequence of the toxic effects indicate that thallium needs to be transported across the cell membrane before it can finally interfere with the release mechanisms. This rather indirect mode of action of thallium was also found in the recordings of presynaptic ion currents.

Perineural recording techniques and the blocking of potassium channels excluded the possibility that presynaptic potassium or calcium channels were influenced by thallium in acute superfusion experiments. Thus, the mechanisms that cause enhancement of the spontaneous release of acetylcholine and the reduction of phasic transmitter release at the neuromuscular junction, both of which are induced by thallium, remain unknown at present (Wiegand et al., 1990).

7.9 *In vitro* test systems: cell lines

Ultrastructural studies with cultured fetal mouse heart cells showed swollen mitochondria with loss of cristae, disintegration of the membrane system, and a protective effect of selenium (Liu, 1986) (section 7.10.2).

A cytotoxic effect on ovary cells was observed in *in vitro* experiments on Chinese hamster ovary cells. After a 16-h incubation with thallium(I) nitrate (40 mg/litre), 50% of the cells did not form colonies during the following 7 days (section 7.9) (Hsie et al., 1984; Tan et al., 1984).

Using three mammalian cell lines (human diploid embryonic fibroblasts, HeLa cells and mouse fibroblasts) to test 11 heavy metals, thallium was found to belong to the group of metals with a strong inhibitory (on proliferation) or lethal effect. After treatment for 7 days, the minimal inhibitory concentrations of thallium(I) and thallium(III) for all three cell lines were 4 mg/litre and 2 mg/litre, respectively, while the 50% inhibitory concentrations were 10-15 mg/litre and 5-15 mg/litre, respectively. Half of the cells were killed by 20-40 mg/litre (Fischer, 1981).

7.10 Factors modifying toxicity

7.10.1 *Enhancement of elimination*

Various substances have been evaluated for their ability to enhance faecal or renal elimination of thallium (e.g., Lund, 1956b). In rats, potent diuretic agents such as furosemide and ethacrynic acid increased renal elimination of thallium, but did not further increase the elimination induced by feeding a potassium-rich diet (Lameijer & van Zwieten, 1977a,b). A sodium-rich diet did not promote the renal elimination of thallium (Lameijer & van Zwieten, 1979). In rats, oral application of activated charcoal increased faecal elimination by about 80% but did not affect urinary elimination (e.g., Lund, 1956b). In contrast, potassium chloride and cystine only increased renal elimination by 47% and 60%, respectively. Meyer & Tal (1957) reported that some compounds containing sulfur and labile methyl groups seem to reduce the toxicity of thallium in rats.

Dithizone (diphenylthiocarbazone), which forms a firm complex with thallium, increased faecal elimination by 33% and urinary elimination by 75% in rats (Lund, 1956b). After treatment with diethyldithiocarbamate "dithiocarb", the resulting lipophilic thallium(I) chelate readily passed the blood-brain barrier and was rapidly decomposed in the brain (Kamerbeek et al., 1971a). However, it did not protect animals from death after administration of lethal doses of thallium (Danilewicz et al., 1980). The benefit of other agents, such as the diuretic 2,3-dimercapto-succinic acid (Liang et al., 1980) or other compounds with mercapto groups (Oehme, 1972), has still to be proven in the treatment of thallotoxicosis.

Comparing different antidotal treatments in rats, Lehmann & Favari (1985) found an increase in thallium elimination to 99% by dithizone, 93% by activated charcoal, 64% by the diuretic agent furosemide, 82% by Prussian Blue, and 92% by combining Prussian Blue and furosemide, whereas the untreated controls had only eliminated 53% of the administered dose (2 mg) within 8 days. After treatment with Prussian Blue and D-penicillamine in combination, the dangerous redistribution of D-penicillamine did not occur and elimination of thallium was better than after administration of Prussian Blue alone (Rios & Monroy-Noyola, 1992). The efficiency of Prussian Blue could be increased by synthesizing batches with a smaller crystal size (Rios et al., 1991; Kravzov et al., 1993). New compounds, such as rhyolith, *N*-acetylcysteine and dimercaprol, were no more, or even less, effective (Dvorák, 1973; Henderson et al., 1985).

7.10.2 Selenium

Selenium not only protected isolated fetal hearts against damage by thallium (section 7.9), but also decreased its lethal effect in young rats; however, thallium-induced hair-loss was not prevented (Ewan, 1978; Ostádalová & Babicky, 1987). Thallium inhibited pulmonary elimination of volatile selenium compounds, increased the retention of selenium in kidneys and liver, and did not protect against chronic selenosis (Levander & Argrett, 1969). Thallium did not affect the distribution of selenium in the body of mice; in *in vitro* systems it seemed to interact with selenite in glutathione solution and in erythrocytes (Naganuma et al., 1983).

7.11 Mechanisms of toxicity - mode of action

Although several (perhaps interconnected) mechanisms have often been postulated, the exact mechanism of thallium toxicity is still unknown (Cavanagh et al., 1974; Prick, 1979; Sabbioni & Manzo, 1980; Nessler & Briese, 1985; Chandler & Scott, 1986; Cavanagh, 1988).

Conflicting results have been obtained with respect to the effects of thallium on sodium/potassium ATPase activity *in vitro* (stimulation) (Ivashchenko & Balmukhanov, 1974) and *in vivo* (inhibition) (Mourelle et al., 1988). The irreversible inhibition of the unidirectional transport of sodium may be due to an inhibition of transport energy (Skulskii & Lapin, 1983). The affinity of the thallium ion for the sodium/potassium ATPase is 9-10 times greater than that of the potassium ion (Britten & Blank, 1968; Inturrisi, 1969). Since the thallium permeability of biological membranes is usually 10 to 100 times greater than that of potassium and since a similar activation of membrane sodium/potassium ATPase in human erythrocytes and rat liver cells is caused by thallium concentrations that are ten times lower than those of potassium and higher concentrations of ouabain are needed to inhibit thallium-activated ATPase, a high selectivity of thallium(I) ions for potassium transport pathways exists (Skulskii et al., 1973, 1975; Gutknecht, 1983; Favari & Mourelle, 1985; Zeiske & van Driessche, 1986). However, *in vitro* studies must take into account the fact that concentrations of only 0.01 g thallium/litre are likely to occur in cells after lethal poisoning with 10 mg/kg body weight, assuming uniform distribution and no elimination (Burger & Starke, 1969).

Owing to the similarity in ionic radii of potassium and thallium, and since the affinity of the thallium ion for sodium/potassium ATPase is greater than that of the potassium ion, thallium accumulates within the cell at the expense of potassium. The strong interaction of thallium with sites normally occupied by potassium may block cycles that depend on recurrent potassium translocation (Sabbioni & Manzo, 1980).

Once inside the cell, various mechanisms are evident, e.g., effects on other enzymes (Sabbioni & Manzo, 1980), the inhibition of protein synthesis (Hultin & Näslund, 1974), the antimitotic effect of thallium compounds (section 7.9) (Sabbioni & Manzo, 1980), especially in the testes (sections 7.5 and 8.5.1), and/or the involvement of riboflavin vitamin B_{12}, which is coenzyme to a number of enzymes (Emsley, 1978; Nessler & Briese, 1985). Any effect on riboflavin or on enzymes containing sulfhydryl groups (see below) should result in a disturbance of pyruvate metabolism (summarized by Nessler & Briese, 1985). Experimental animals suffering from riboflavin deficiency show symptoms similar to those of thallium intoxication (Schoental & Cavanagh, 1977).

Another postulated mechanism considers the general capacity of thallium to react with thiol groups, thus interfering with a variety of processes (Zeiske & van Driessche, 1986). Although the toxic effect of thallium is reduced by diets high in cystine, methionine and betaine (Oehme, 1972), interference with the metabolism of sulfur-containing amino acids does not seem to be directly involved in toxicity (Garcia Bugarin et al., 1989) and strong reactions with thiols were observed for thallium(III) but not for thallium(I) compounds (Douglas et al., 1990). An additional aspect of the reaction of thallium with sulfhydryl groups is the induction of free radical formation (section 7.8). This is indicated by increased lipid peroxidation rates in the brain. Processes leading to lipid peroxidation finally damage cell membranes by subsequent reactions of free radicals with sulfhydryl enzymes (section 7.11). Lipid peroxidation results in a deficiency of glutathione and leads to the accumulation of lipid peroxides in the brain, liver and kidney and, presumably, finally to lipofuscin granules (Herman & Bensch, 1967; Aoyama et al., 1988). The possibility that this mechanism is responsible for the neurotoxic effects of thallium was supported by the simultaneous administration of thallium and acetyl-homocysteine thiolactone, which prevented reduction in the level of sulfhydryl radicals in the cerebellum and significantly increased glutathione levels (Hasan & Haider, 1989). This hypothesis is further supported by the protective effect of 1) silymarin against thallium hepatotoxicity (Mourelle et al., 1988), 2) selenium, demonstrated with the thallium-induced ultrastructural changes found in isolated fetal mouse heart cell (sections 7.9 and 7.10.2.) (Liu, 1986), and 3) selenium and vitamin E against membrane damage by uncontrolled lipid peroxidation *in vivo* (Hasan & Ali, 1981).

The mode of action of thallium seems to be mainly based on a disturbance in the function of the mitochondria (Barckow & Jenss, 1976), although they are affected by high concentrations of almost all heavy metals (Byczkowski & Sorenson, 1984). The thallium(I) cation may either enter isolated rat liver mitochondria passively, i.e. in an energy-independent manner (Barrera & Gómez-Puyou, 1975), or penetrate the mitochondrial membrane electrophoretically (Skulskii et al., 1978). The entry of thallium into the intramitochondrial space and the interaction of cytosolic thallium with mitochondria membranes may explain the deleterious effects (Skulskii et al., 1984). In isolated mitochondria, thallium(I) acetate caused an uncoupling of oxidative phosphorylation and swelling of the isolated mitochondria (Melnick et al., 1976). Using ascite tumour cells *in vitro*, Ivashchenko et al. (1973) found a strong, thallium-induced increase in oxygen consumption and lactic acid production, which were inhibited by ouabain and sodium fluoride. However, whereas oxygen consumption and anaerobic glycolysis of tissues were affected *in vitro*, tissues from rats with chronic or acute (first day) poisoning did not differ from those of controls (Thyresson, 1950).

Comparing the effects of monovalent and trivalent thallium on isolated rat liver mitochondria, only thallium(III) nitrate uncoupled oxidation from phosphorylation (Hollunger, 1960). This effect could not be reversed by adding edetic acid or dimercaprol (sections 7.10.1 and 8.6).

8. EFFECTS ON HUMANS

The toxicology of thallium is summarized in Fig. 1.

8.1 General population exposure

Thallium concentration in early-morning urine samples of nine non-exposed subjects ranged from 0.13 to 1.69 μg/litre (Weinig & Zink, 1967). Smith & Carson (1977) gave a range of 0.6 to 2.0 μg/litre with a mean urinary thallium concentration of 1.3 μg/litre.

The initial use of thallium in medicine, mainly as a depilatory drug, caused many cases of intoxication. The observation of side effects led to more detailed toxicological studies after 1918, and its medical use was abandoned after 1945 (Emsley, 1978; Briese & Nessler, 1985a). Since thallium is tasteless, odourless, without colour and highly toxic, and used to be easily obtainable, it was often used for suicide, homicide and attempts at illegal abortion (Kemper & Bertram, 1984; Manzo & Sabbioni, 1988). Also large numbers of accidental intoxications have occurred, e.g., in Berlin, Germany there were 110 cases of accidental ingestion and 39 attempted suicides by children involving corn poisoned with thallium sulfate from 1967 to 1976; three of the children died (von Mühlendahl et al., 1978). Munch (1934b) summarized thallium intoxications prior to 1934. He found reports on 8006 children who had been treated with thallium as a depilatory agent. Intoxication occurred in 447 cases and 8 children died. In connection with its use as a rodenticide or insecticide, 21 poisonings and 5 deaths occurred.

Recently, intoxications occurred after ingestion of a Chinese herbal medication/nutritional supplement containing 30 g thallium per litre, but other samples contained no thallium (Schaumburg et al., 1992).

8.1.1 Acute toxicity

Cases of acute intoxication by thallium salts in humans, which always cause severe symptoms, have been reported for single or multiple oral doses of the order of 100 mg or more for adults, i.e. 1.5 to 2 mg/kg body weight (Table 25).

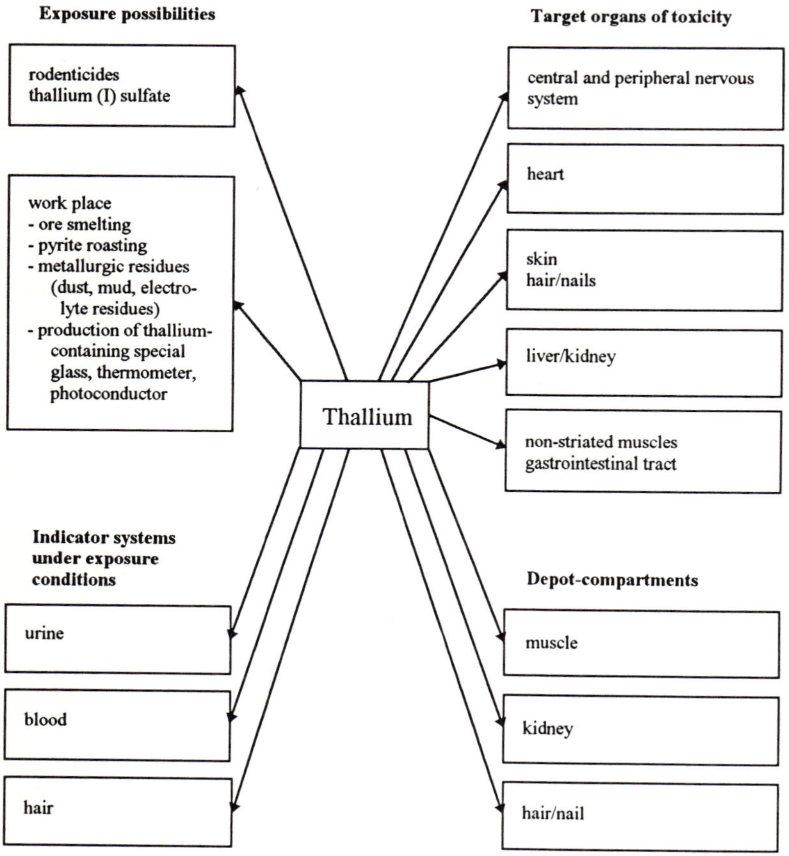

Fig. 1. Toxicology of thallium (modified from Bertram & Kemper, 1983)

Symptoms of acute thallium toxicity vary with age, dose and route of administration (Venugopal & Luckey, 1978). According to Sharma et al. (1986), one-tenth (2-10 mg/person) of the lethal thallium dose for adults causes death in children. However, reports of the therapeutic use of thallium in which children tolerated larger doses than adults indicate the contrary (Ormerod, 1928;

Table 25. Acute toxicity of thallium

Toxicity[a]	Dose	Adult/children	Reference
TT	> 1.5 mg/kg	not specified	Schoer (1984)
MLD	10-12 mg/kg 2-10 mg/kg	adult children	Kazantzis (1994)
MLD	14-18 mg/kg	not specified	Sessions & Goren (1947)
Lethality	20-100 mg per person 2-10 mg per person	adult children	Sharma et al. (1986)

[a] MLD = minimal lethal dose; TT = toxicity threshold

Sessions & Goren, 1947; Prick, 1979). Of 75 children who had accidentally ingested thallium sulfate, only two showed symptoms of thallium intoxication: a 2-year-old child who had ingested 3.5 mg thallium/kg and eliminated 130 μg thallium/litre in the urine showed no reflex movements of the legs in the second week, while another child with 3.9 mg thallium/litre in its urine and 40 μg/litre in its blood showed no neurological symptoms apart from a massive alopecia in the second week. In one of two lethal cases, the urine contained 10.3 mg/litre, and the death of both children occurred within 2 days (von Mühlendahl et al., 1978). The influence of age is unclear, but this may be due to the reporting of doses either as total dose or as mg/kg body weight. Therapeutic uses of thallium in the 1930s (8 mg/kg) resulted in peripheral neuritis in about 10% of the patients (Cavanagh, 1979).

In adults lethal doses vary between 6 and 40 mg/kg (500 to 3000 mg/person), with an average dose of 10-15 mg/kg body weight (Schoer, 1984). Without therapy the average dose usually results in death within 10-12 days (Kemper, 1979), but, summarizing 150 mostly suicidal cases of thallium intoxication, most of the patients who had ingested about 600 mg and had no gastric lavage within the initial 2 h died within 8-10 h (Potes-Gutierrez & Del Real, 1966). After accidental ingestion of 10-fold overdoses of thallium acetate (80 mg/kg), given as a depilatory in ringworm disease (Tinea), four children died after 1-2 days (Lynch et al., 1930). Sessions & Goren (1947) had previously suggested 14 to 18 mg thallium/kg body weight to be the minimal fatal dose.

The triad of gastroenteritis, polyneuropathy and alopecia is regarded as the classic syndrome of thallium poisoning (Gastel, 1978), but in some cases gastroenteritis and alopecia are not observed. Other symptoms also develop in varying sequence. Both lethal and sublethal doses give rise to most of the symptoms, but these same symptoms vary in intensity and time, probably in a dose-dependent way. The following general trends have been summarized from numerous reviews and case reports, e.g., Muller (1961), Moeschlin (1965, 1986), Potes-Gutierrez & Del Real (1966), Venugopal & Luckey (1978), Gastel (1978), Möllhoff et al. (1979), Sabbioni & Manzo (1980), Davis et al. (1981), Saddique & Peterson (1983), Kemper & Bertram (1984), Le Quesne (1984), Schoer (1984), Ohnesorge (1985), Nessler & Briese (1985), Chandler & Scott (1986), Arnold (1986), Hayes & Laws (1991) and ATSDR (1992).

People who died within 8-10 h showed increasing tachycardia, progressive hypotension, early hyporeflexia and peripheral cyanosis. The ingestion of lower lethal doses causes gastrointestinal haemorrhages (blood in faeces), gastroenteritis, metallic taste, salivation, nausea and vomiting. Neurological disorders become apparent within 2-5 days irrespective of the route of administration. Within 5-7 days hallucination, lethargy, delirium, convulsions, a tingling pain in the extremities and muscular weakness are followed by coma. The cause of death is respiratory failure or cardiac arrest.

The sequence of symptoms in less severe intoxications is outlined in Fig. 2. Similar clinical symptoms develop after ingestion of lethal doses, which are promptly treated by enhancing elimination (Graben et al., 1978). Within hours after ingestion, thallium often induces nausea or vomiting, which may also appear in the next 2 days. Other initial symptoms, e.g., diarrhoea, abdominal pain and a dull feeling in body extremities, occasionally occur. Constipation is common and may be difficult to treat, thus interfering with antidotal treatment with Prussian Blue (section 8.1). Starting at day 4, a dark region, resembling melanin pigment, may appear in the hair roots (this could be of diagnostic value).

Following this latent period, in the early phase lasting about 1 week, some of the typical thallium disorders slowly develop

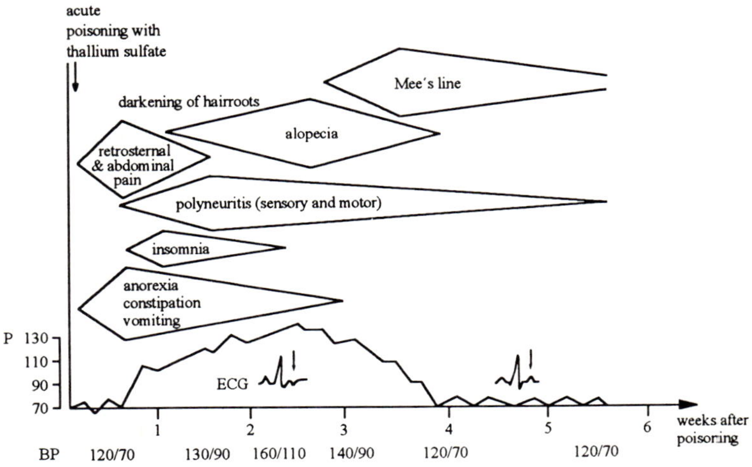

Fig. 2. Typical time course of acute thallium poisoning in one case with recovery (modified from Moeschlin, 1986)
P = pulse rate; BP = blood pressure

(culminating in the third or fourth week). Firstly, retrosternal and abdominal colic-like pains, as well as pain and tenderness in the legs, often become prominent. Excessive thirst, sleeplessness, restlessness, hysteriform behaviour and electro-encephalographic abnormalities indicate involvement of the central nervous system. A characteristic symptom of sensory neuropathy is the extreme sensitivity of the lower extremities. The neurological syndrome can also include optic neuritis, numbness of fingers and toes with loss of sensation to pin-prick and touch, and the "burning feet syndrome". As an additional sign of a mixed sensory-motor neuropathy, ankle reflexes are lost early, while other reflexes may be maintained for a time or even increased. During this phase, thallium intoxication can mimic a systemic lupus erythematosus or a pseudobulbar paralysis (Guillain-Barré syndrome or Landry's ascending paralysis) (Gastel, 1978; Alarcón-Segovia et al., 1989; Cavanagh, 1991). The renal function is generally not affected in the early course of poisoning; only a slight albuminuria with formed elements in the urine may be found. Urinary elimination of porphyrins and porphyrin precursors may be greatly increased during this early phase (Merguet et al., 1969; Paulson et al., 1972; Bank et al., 1972; Graben et al., 1978).

During the second week hypertension and tachycardia are frequently observed symptoms (Romero Romero et al., 1989). Sometimes peroneal paralysis and atrophy of other muscles may develop. After a short phase of perspiration, the skin becomes dry and scaly (probably due to an effect on the sweat and sebaceous glands) and sometimes necrotic. Damage to hair papillae seems to be responsible for loss of head hair. This frequently begins during the second week. Complete depilation occurs within about one month and regrowth begins some time later, often without any pigment.

About 3 to 4 weeks after poisoning, dystrophy of the nails is shown by the appearance of white lunular stripes (Mees's stripes), which are also observed in cases of arsenic poisoning (Buschke & Langer, 1927; Greving & Gagel, 1928).

After 4 to 5 weeks, survival of the patient is likely, but recovery requires months. Sometimes neurological and mental disturbances, as well as electro-encephalographic abnormalities and, rarely, forms of paranoia, persist. Occasionally, cataract (opacity of the eye lens) has been described. In children a high percentage of the neurological disorders were still present after 4 years. Double optic atrophy in one patient after 3 months was reported (Munch, 1934).

8.1.2 *Effects of long-term exposure: chronic toxicity*

Studies of long-term exposure to thallium resulting in chronic poisoning have been summarized by Buschke & Langer (1927), Moeschlin (1965), Gefel et al. (1970), Schoer (1984) and Goldblatt (1989) without any information about doses. The symptoms show strong variation and are in general milder than in cases of acute intoxication. Depending on the level of exposure, a relatively long latent period (several weeks) may be followed by just a few symptoms. Peripheral sensorial disturbances, mental aberrations, loss of weight and sleeplessness seem to be the most common (Valentin et al., 1971; Sabbioni & Manzo, 1980; Nessler, 1985b). In more severe cases, disturbances of vision, pain without marked polyneuritis, and loss of hair were reported. Later, severe polyneuritis may develop, with an inability to walk, amaurosis (blindness) and pronounced cachexia. Cardiac disorders include hypertension, irregular pulse and angina-like pain. Renal dysfunction is indicated by albuminuria and haematuria. Other symptoms are gastric anacidity, lack of appetite, loss of weight, endocrine disorders, psychoses and encephalitis.

Complete rehabilitation takes months and can be interrupted by relapses, probably caused by remobilization of thallium from tissue depots.

Epidemiological studies carried out in the contaminated area of Lengerich, Germany, comprising about 1200 people, revealed positive correlations between the concentration of thallium found in urine or hair samples and polyneuritic symptoms such as paraesthesias and pain in muscles and joints, as well as psychasthenic symptoms such as headache, sleep disorders and fatigue. No correlation was found with respect to gastrointestinal troubles or skin disorders. Surprisingly, a negative correlation with hair loss was found. Only one of 51 people with > 20 μg thallium/litre urine showed lunular stripes in the nails (LIS, 1980; Brockhaus et al., 1980, 1981b; Dolgner & Wiegand, 1982; Schoer, 1984). Strong individual variation in sensitivity prevents an estimation of the thallium concentration in the urine at which no effects occur (Dolgner & Wiegand, 1982).

8.2 Occupational exposure

There have been numerous reports of factory workers with thallium poisoning, but no fatal cases have occurred. Peripheral sensorial disturbances, mental changes, loss of weight, and sleeplessness are the symptoms which seem to prevail (Munch, 1934b; Muller, 1961; Malcolm, 1979; Saddique & Peterson, 1983; Triebig & Büttner, 1983; Schoer, 1984; Nessler, 1985b; Junghans & Nessler, 1985; Ohnesorge, 1985; Kazantzis, 1986). In Germany, the United Kingdom and some other countries, thallium poisoning represents an occupational disease entitling the victim to compensation. In Germany, such compensation was granted in three cases between 1970 and 1985 (Ewers, 1988).

Increased thallium concentrations in the urine of workers have often been found. For example, the urine of workers in a company producing alloy anode plates for use in magnesium sea water batteries contained up to 236 μg thallium/litre, but no differences in medical records of exposed and unexposed workers could be demonstrated (Marcus, 1985). In his review of thallium, Ohnesorge (1985) summarized several reports of industrial poisoning. Exposure over several months or years resulted in typical thallium symptoms, e.g., leg pains, tiredness, alopecia and psychological disorders, but also (in one case) blindness. Permanent blindness was also reported in another review by McDonald (1941). Exposure to more than 0.01 mg thallium/m^3 for 16 to 17 years caused disorders of the vascular system, as well as neurological symptoms (Ohnesorge, 1985). From the triad of gastroenteritis, polyneuropathy and alopecia, only disorders of the gastrointestinal tract were not reported.

Glomme (1983) emphasized that objective symptoms of polyneuritis may not be demonstrable for some time. In addition to the changes in the superficially provoked tendon reflexes, a pronounced weakness and a fall-off in the speed of pupillary reflexes can occur.

In a further study on cement plant workers, 36, selected at random, were subjected to thallium analyses of blood, urine and hair, together with a neurological examination and electrophysiological investigation including sensory and motor nerve conductive velocities, evoked potentials and electro-encephalography (Ludolph et al., 1986). One half of the workers examined suffered from concurrent disorders, including diabetes mellitus. Although multiple symptoms and signs of neurological disorders were detected, no correlation was found between the electrophysiological findings and thallium levels in blood, urine and hair. Urinary thallium levels were above 5 μg/litre in five of the examined workers. Blood thallium levels above 2 μg/litre were found in 16 workers and hair thallium levels above 20 μg/kg in four workers. The investigators concluded that more thorough epidemiological techniques would be required to reveal a possible causal relationship between chronic low-dose thallium exposure and neurological deficits.

8.3 Subpopulations at special risk

There are no subpopulations at special risk of thallium intoxication except workers in the respective industries and populations living in thallium-contaminated areas. There are no good data to suggest that infants or pregnant women are more sensitive to the effects of thallium than the general population. The available data, however, are inadequate to fully assess these subpopulations. Because thallium is eliminated in both urine and faeces, any subpopulations with diminished excretory capabilities (e.g., renal insufficiency) may be at increased risk of thallium poisoning. It has been recommended that workers be excluded from working with thallium if they suffer from renal or hepatic disease, anaemia, blood dyscrasias, hypertension, alcoholism, chronic infections or endocrine gland dysfunction. It has also been recommended that workers potentially exposed to thallium should be encouraged to eat potassium-rich food, as thallium and potassium ions can mimic each other *in vivo*. Accordingly, potassium-deficient individuals may also be at increased risk from thallium toxicity.

8.4 Target organs in intoxicated humans: pathomorphology and pathophysiology

Effects on the different organs have been summarized by Prick (1979), Sabbioni & Manzo (1980) and ATSDR (1992). In nearly all affected organs direct cytotoxic effects, as well as indirect effects, caused by damage to the nervous system, have been found (Prick, 1979).

8.4.1 Gastrointestinal tract and renal system

In a fatal case of thallium poisoning, in which the woman died after at least 14 days, there was gross dilatation of the stomach and a thin "blue line" was evident at the margin of the gingiva of the lower incisors, but no alterations of the intestinal wall were apparent (Curry et al., 1969). Other patients who died 1 to 16 days after oral poisoning showed hyperaemia, congestion of the gut, punctate haemorrhages in the mucosa of the stomach and upper intestinal tract, and swelling of the mucosal cells (Lynch et al., 1930; Munch et al., 1933; Heath et al., 1983). As a result of depilatory treatment in children, gastric hypoacidity was reported (Buschke, 1929), an effect also observed after a suicide attempt (Greving & Gagel, 1928).

In several cases of oral poisoning, usually fatal, the liver was usually found to be congested, greyish yellow or yellow in colour, had microscopic fatty infiltrations of the hepatocytes and a tendency to central necrosis (Lynch et al., 1930; Munch et al., 1933; Curry et al., 1969).

At least 6 weeks after intoxication, renal biopsy of a patient with 13.8 mg thallium/litre in his urine showed diffuse proliferative glomerulonephritis with granular immunofluorescence for IgG, IgM and C3 (Alarcón-Segovia et al., 1989). In the postmortem examinations by Lynch et al. (1930), Munch et al.

(1933) and Curry et al. (1969), sections of kidney were dull red or congested and showed marked hyperaemia, cloudy swelling of tubules and degenerative changes of glomeruli. Weinig & Schmidt (1966) also reported kidney damage (but perhaps from a previous attempt at poisoning) in a woman and her son who died after taking thallium. This kidney damage may have been responsible for the relatively low thallium concentrations in the son's kidney in comparison to concentrations in other tissues (section 6.2.2).

8.4.2 Cardiovascular system

Accidental poisoning of three children, who died within two days, caused fatty degenerations in the victims' hearts. These were more marked and more dispersed in the youngest child, who survived longest (Lynch et al., 1930). In some cases of postmortem-diagnosed thallium poisoning, resulting in death within 4-14 days, fresh haemorrhagic myocardial lesions were found (Heath et al., 1983; Andersen, 1984), while in another case only a few focal haemorrhages were present (Curry et al., 1969). Haematological changes, e.g., anaemia, leucocytosis, eosinophilia, thrombocytopenia (at least partly resulting from a toxic effect on bone marrow) and lymphopenia, have been summarized by Saddique & Peterson (1983) and Luckit et al. (1990).

In five patients suffering from severe and protracted thallium poisoning, cardiovascular changes were recorded (Machtey & Bandmann, 1961; Franke et al., 1979). The patients' blood pressure showed marked fluctuations, even in the course of one day, but systolic and diastolic hypertension occurred only on a temporary basis. The authors believed that these changes and also the observed tachycardia and electro-cardiographic changes were caused by direct effects of thallium on the autonomic nervous system. Tachycardia can appear about a week after intoxication and last for 5 weeks (Franke et al., 1979). Involvement of the autonomic nervous system is also indicated by changes in renal function and by the urinary concentrations of various metabolites (brenzcatecholamines, vanillin-mandelic acid, β-aminolaevulinic acid, porphobilinogen, coproporphyrin and total porphyrins) during hypertension and tachycardia resulting from thallium poisoning (Bock et al., 1968). Concentrations of brenzcatechinamines and porphobilinogen were greatly increased, and hypertension and tachycardia could be influenced by administration of α- and β-receptor blockers. In addition, increased elimination of brenzcatecholamines, which presumably originate not only from the adrenal medulla but also from the sympathetic nervous system, indicates a strong stimulation of the adrenergic system (Bock et al., 1968).

8.4.3 Skin and hair

Five young men suffering from thallium poisoning showed follicular plugging of the skin on the nose and cheeks and in the nasolabial folds by keratinous material, crusted eczematous lesions and acneiform eruptions on the face, dry scaling on palms and soles, and alopecia, not only of the scalp but sometimes also of the eyelashes, lateral eyebrows, arms and legs. Histological examination of skin biopsies from both scalp and cheek showed disintegrating hairshafts, gross follicular plugging and eosinophilic keratohyaline granules in the adjacent granular layer of the epidermis. Sebaceous glands were sometimes necrotic. Biopsies of the pustular lesions on the face showed folliculitis and necrosis of the follicles, while in those from the feet marked hyperkeratosis and hypergranulosis were evident (Heyl & Barlow, 1989). Effects on the follicles are also reported by Hausman & Wilson (1964) and Bonnet & Pedace (1979), but in a woman who died at least 14 days after intoxication no hyperkeratosis in any part of the skin was found (Curry et al., 1969).

As is the case in experimental animals (section 7.4.1), the reason for the different sensitivities of different types of hair (lanugo, pubic and axillary hair is much less or is later affected than hair of the head) in humans is unclear (Buschke & Peiser, 1926; Buschke, 1929; Cavanagh, 1988). Cavanagh et al. (1974) emphasized a direct effect on the keratinocytes, and Cavanagh (1988) finally suggested that the difference is due to the fact that hair follicle cells are only affected when they are mitotically active.

The depilatory effect generally does not result in permanent hair loss. Since the new hairs which grow following thallium-induced alopecia are stronger than those lost and also develop in regions which had been hairless prior to the thallium poisoning, Buschke successfully used thallium in therapy of alopecia induced by hair disease (Buschke & Curth, 1928).

Soon after poisoning, hair papilla are seen to contain black regions and the growing end is tapered (e.g., Hausman & Wilson, 1964; Curry et al., 1969; Saddique & Peterson, 1983). In some reviews this phenomenon is interpreted as black pigmentation. However, Ludwig (1961) had already shown that these regions do not contain deposits of pigments or thallium but small amounts of air which had entered the shaft. Later investigations demonstrated that the gaseous inclusions result from a trophic disorder in keratin formation (Kijewski, 1984; Metter & Vock, 1984). In both investigations, scanning electron microscopy demonstrated a loosening of the elements of the fibre layer of the hairs.

8.4.4 Nervous system

Neurological disorders showing strong variability are one of the three major symptoms of thallium poisoning (Möllhoff et al., 1979; Prick, 1979; Sabbioni & Manzo, 1980; Le Quesne, 1984; Manzo & Sabbioni, 1988). In contrast to the other disorders, neurological deficits usually persist. Ataxia, mild spastic paraparesis and impairment of intellectual powers developed after treatment of scalp ringworm with thallium and persisted for 36 years, and it is possible that increasing problems with mobility after 33 years were also due to the treatment (Barnes et al., 1984).

8.4.4.1 Central nervous system

Like some other heavy metal intoxications, those caused by thallium are usually associated with subacute and chronic (but rarely with acute) encephalopathy (Rosenstock & Cullen, 1986). In one patient who died 9 days after ingesting 5-10 g thallium nitrate, no abnormalities were evident in histological or ultrastructural examinations of the central nervous system (Davis et al., 1981). The brain of another patient, who died within 4 days of intoxication and had an extreme postmortem concentration of 36 mg/litre in his blood, was moderately swollen (Andersen, 1984).

Seven people who died 11 to 16 days after accidental ingestion of thallium had localized oedema and various grades of chromatolysis in their neurons, especially those of the pyramidal tract, the third nucleus, the substantia nigra and the pyramidal cells of the globus pallidus. Blood vessels were distended with blood (Munch et al., 1933).

In a fatal case of thallium poisoning, the brain of the dead person was slightly swollen and oedematous about 4 weeks after the ingestion of around 33 mg thallium sulfate/kg body weight. Petechial haemorrhages were found in the white matter, particularly in the parietal regions and subthalamic areas. The brain stem and cerebellum showed a normal appearance. Axonal swelling and fragmentation in the cortico-spinal tracts could be traced through the mid-brain, pons and medulla into the spinal cords. Chromatolysis of brain stem nuclei was only marked in facial and hypoglossal nuclei and nerve fibre degeneration only in the spinal tract of the 5th nerve (Kennedy & Cavanagh, 1976).

In a suicidal case, general degeneration of ganglion cells, damage to axons and disintegration of myelin sheaths were observed in the brain of the person, who died 21 days after intoxication. Fatty degeneration of ganglion cells, acute swelling of oligodendroglia, a spongy appearance of the basal ganglia and a particular concentration of lesions in the calcarine cortex were prominent (Karkos, 1971). In another fatal case, autopsy showed degeneration of ganglion cells in the brain and spinal cord (Gefel et al., 1970).

Detrimental effects on intellectual functions were assumed not only in a patient suffering from ringworm treatment (section 8.4.4) (Barnes et al., 1984), but also in a student of chemistry who eliminated 60 mg thallium/litre in his urine after poisoning in the laboratory (Thompson et al., 1988). The data obtained from intelligence tests on the student, performed 7 and 13 months after the near fatal intoxication, were compared with those of his non-identical twin brother who was of a similar educational background. Although the brothers are not totally comparable, the tests indicated severe deterioration particularly in memory and performance abilities and, 13 months later, there was only little general improvement.

.4.4.2 *Peripheral nervous system*

Histological and ultrastructural examination of postmortem samples can produce inconsistent results, presumably because of the different periods of time between intoxication and sampling and because of differences in dose size. In general, clinical symptoms and signs can be correlated to neuropathological findings (Cavanagh, 1979).

Damage to the autonomic nervous system accounts for many of the effects on various organs, e.g., fever, tachycardia, labile blood pressure, orthostatic hypotension, urinary retention, constipation and cardiac arrhythmias (Gastel, 1978; Prick, 1979). Thallium intoxication causes symmetric, mixed peripheral neuropathy (Rosenstock & Cullen, 1986). Distal nerves are more strongly affected than more proximal nerves, and earlier but lesser degrees of change occur in nerves with shorter axons, e.g., the cranial nerves (Cavanagh et al., 1974; Cavanagh, 1979, 1988).

a) Histology and ultrastructure

Neuropathological findings vary. Little evidence of neuronal degeneration in the sciatic nerve or spinal cord were found in a person who died about 14 days after intoxication (Cavanagh et al., 1974). In a fatal case of poisoning, in which the patient died just 9 days after intoxication (section 8.4.4.1), a sural nerve biopsy was obtained 2 days before death. In addition, postmortem samples of peripheral and cranial nerves and sections from various parts of the central nervous system were taken. Ultrastructural examination of the sural nerve showed that the myelin sheaths had often disintegrated into a series of ovoids along the course of the axon (Davis et al., 1981). Similar findings have also been reported from other sural nerve biopsies, taken, for example, 3 days in one case and at least 5 to 6 weeks after thallium poisoning, from patients who survived (Bank et al., 1972; Paulson et al., 1972; Alarcón-Segovia et al., 1989; Dumitru & Kalantri, 1990). Degenerated myelin sheaths contained myelin figures and electron-dense granules, whereas axons usually had a normal appearance and rarely contained densely packed neurofilaments (Bank et al., 1972). Munch et al. (1933) and Davis et al. (1981) found axon degeneration in peripheral nerves, even in axons with ultrastructurally normal myelin sheaths; axons were swollen and contained vacuoles and distended mitochondria. Non-myelinated axons on the other hand showed only slight or no abnormalities. Beadings of axons were not only present in distal portions of peripheral nerves, but also in some cranial nerves, whereas the other cranial nerves and the proximal portions of peripheral nerves were histologically normal.

In another postmortem examination of a patient who died about 4 weeks after intoxication (section 8.4.4.1) (Kennedy & Cavanagh, 1976), the nerve fibres of several peripheral nerves were severely reduced, long fibres in particular being more severely affected. Changes in neurons of the spinal cord were evident in all regions but most strikingly in the lumbosacral region, where many neurons clearly showed the classical chromatolytic changes which indicate attempted regeneration. Dorsal column changes in the spinal tracts could clearly be correlated in time with the peripheral nerve symptoms, and the slight changes in the lateral cortico-spinal tracts could be traced to the recent necrotic lesions in the diencephalon (Kennedy & Cavanagh, 1976). Demyelination of the dorsal columns in sections of cervical spinal cord was also observed during the postmortem examination of a women who died at least 14 days after intoxication (Curry et al., 1969).

The severe damage to the vagus, denervation of the carotid sinus, and lesions of the sympathetic ganglia found in postmortem examinations indicate the involvement of the autonomic nervous system (Gastel, 1978).

b) Electrophysiological investigations

In a case of thallium poisoning in which the patient survived, a sural nerve biopsy was obtained and nerve conduction and serial electromyographic studies were carried out, beginning 10 days after onset of the symptoms and ending 24 months later (Dumitru & Kalantri, 1990). Initially, the plantar nerves of the foot showed profound axonal loss, from which there was no recovery, as shown by conduction studies over the next 2 years. During the initial 4 months, sural and peroneal nerves also underwent axonal loss but recovered within 2 years. In other cases of thallium intoxication, nerve conduction studies gave normal results or revealed retarded latencies of nerves of the upper (more than the lower) extremities, as well as temporal dispersion indicating demyelination (Alarcón-Segovia et al., 1989).

Sensory fibres of the nervus medianus were examined in a patient with acute thallium poisoning in order to assess the effects on the conduction velocities of faster and slower nerve fibres. Two months after the onset of symptoms the patient showed evidence of distal sensorimotor neuropathy, but only the conduction velocities of faster fibres were below the normal lower limit. Nine months later, symptoms had almost disappeared and conduction velocities of both slower and faster fibres were within the normal range (Yokoyama et al., 1990). Only a slight electrophysiological correlation with the symptoms of a persistent polyneuropathy were reported from an examination carried out 3 years after a case of intoxication (Feudell, 1982).

c) Visual disorders

Retrobulbar neuritis and resulting visual impairment can develop or persist months after termination of treatment with thallium-containing depilatories, and even optic atrophy may occur (e.g., Buschke & Langer, 1927; Lillie & Parker, 1932; Mahoney, 1933; Bank et al., 1972; Bahiga et al., 1978; Tabandeh et al., 1978; Schmidbauer & Klingler, 1979). An ascendent (retinal) atrophy of the optic nerve may result from the toxic effects of thallium on the retina (Hennekes, 1983). Nerve fibres in oculomotor muscles can also show degenerative changes

(Cavanagh et al., 1974). In patients with optic neuritis some reduction in visual acuity always persists (Goldblatt, 1989). About 10 months after thallium intoxication, a keratoconjunctivitis sicca was found to have developed (Alarcón-Segovia et al., 1989).

8.4.5 Other organs

Effects on the lung and endocrine glands were found in six postmortem examinations (death occurred 11 or 15/16 days after ingestion of thallium). Light microscopy showed the alveoli distended with serum, marked hyperaemia and a few areas with bronchopneumonia (Munch et al., 1933). In another fatal case the pleurae were free from haemorrhages and adhesions (Curry et al., 1969). Of the endocrine glands only the adrenals were affected. They showed marked hyperaemia, small haemorrhages in the medulla, areas of necrosis and nuclear disintegration (Munch et al., 1933). In other lethal cases the adrenals were enlarged but without haemorrhages (Curry et al., 1969), or were haemorrhagic (Gefel et al., 1970), or the concentration of lipoids was reduced (Buschke, 1929).

A biopsy, taken 50 days after intoxication, showed marked areas of atrophy of muscle tissue (Franke et al., 1979). Muscle fibrosis was reported by Gefel et al. (1970) in a fatal case of thallium poisoning.

8.5 Special effects

8.5.1 Reproduction and developmental effects

Few data are available with respect to the effects of thallium on human reproduction (Schardein & Keller, 1989). Female cycles are arrested, and libido and potency of males decrease (Buschke & Langer, 1927; Greving & Gagel, 1928). Effects on sperm are known to occur in cases of chronic intoxication (Cottier, 1980). It should be noted that minor amounts of thallium accumulate in the testis after diagnostic scintigraphy, but possible effects have not been investigated.

There are no reports of any teratogenic effects in humans and an extrapolation of animal data to humans is somewhat problematical (Kolb Meyers, 1983; Mottet, 1985).

Reviews of more than 20 cases of thallium intoxication during pregnancy by Petersohn (1960), Moeschlin (1965), Stevens &

Barbier (1976), Graben et al. (1980) and Barlow & Sullivan (1982) can be summarized as follows: all attempts at illegal abortion were in vain; the prolonged use of a depilatory cream seems to have been the cause of 1 neonatal death. Two attempts at illegal abortion with thallium in the first trimester of pregnancy did not affect the development of the fetuses, although rather low birth weights were recorded. In four additional cases of intoxication during this period of pregnancy, the outcome was not reported.

Intoxication occurring after the first trimester can induce in the newborn baby some symptoms of acute intoxication seen in adults, e.g., rash and alopecia. Two babies born after the intoxication of their mothers in the 5th and 6th months of pregnancy showed reduced weight or no effects, respectively. Also no effects were found in a case of intoxication (0.35 g thallium) in the 7th month of pregnancy or in an additional six cases. However, in two cases during this period (0.15 g thallium in one case), premature births occurred, showing alopecia areata and low weight of one baby. Alopecia areata and lunular stripes in the nails were observed in two newborn babies. Low birth weight was common.

Petersohn (1960) reported an attempt at illegal abortion by ingesting 0.5 g thallium 8 weeks before term. However, the fetus developed normally. The child had well-developed hair and apart from being relatively underweight showed no signs of thallium poisoning, whereas the mother developed alopecia and polyneuritis (Erbslöh, 1960). A suicide attempt with about 1.2 g thallium 2 days before birth caused the death of a newborn girl after 5 days; fresh blood in the faeces was observed from the 3rd day onwards.

In the population living around the cement plant in Lengerich, Germany, 300 women gave birth in the years 1978 and 1979. Eleven children exhibited congenital malformations or abnormalities, five showing major malformations (e.g., cleft lip and palate, clubfoot, hip dislocation and ventricular septum defect). The rate of malformation was higher than expected, but the authors suggested that the real frequency of malformation in unaffected populations is underestimated. It was difficult to correlate the effects with the intensity of exposure, since the degree of exposure to which the mothers were subjected during pregnancy could not be ascertained (Dolgner et al., 1983). It should be noted that the fathers were not included in the investigations. Embryotoxic effects were not considered in the investigation at Lengerich (Claussen et al., 1981).

8.5.2 Carcinogenicity

The carcinogenicity of thallium has not been adequately evaluated in humans. A study by Marcus (1985) on occupationally exposed workers showed that the incidence of benign neoplasms (not further characterized) was not significantly increased in the workers. However, only 86 thallium-exposed and 79 controls were included in this study and the length of observation time was not stated. The study was also limited by the availability of medical records. Other reports involving human exposure to thallium did not include an investigation of carcinogenicity.

8.5.3. Immunotoxicological effects

Reduced resistance against secondary infections has been reported only by Moeschlin (1965), but actual data on the possible immunological effects of thallium are not available.

8.6 Factors modifying toxicity: enhancement of elimination

In studies on laboratory mammals (section 7.10.1) and in tests with patients, enhancement of elimination was attempted (Stevens et al., 1974). This might be achieved, provided that the thallium is not fixed intracellularly (Barckow & Jenss, 1976). Sodium salts were previously used as an antidote for human thallotoxicosis (Munch, 1934a), but intravenous injection of sodium thiosulfate (Sessions & Goren, 1947) often increased the severity of symptoms (Munch, 1934a). Although increased urinary elimination of thallium theoretically should reduce its fatal effects, treatment with potassium salts caused a worsening of the symptoms of thallotoxicosis in humans (Papp et al., 1969). This was presumably due to a mobilization of intracellular thallium, an increase in plasma levels, and redistribution (Bank et al., 1972; Gastel, 1978).

Dithizone has also been used to treat cases of human poisoning (summarized by Bendl, 1969 and Papp et al., 1969), in spite of its goitrogenic and perhaps diabetogenic effects in experimental animals. Clinical therapy with dithizone is often more effective than treatment with potassium chloride and charcoal (Paulson et al., 1972). Respiratory distress, confusion and diplopia have been cited as examples of negative side effects by Barckow & Jenss (1976) and were also reported by Saddique & Peterson (1983), but they were not attributed to the dithizone treatment by Paulson et al. (1972). Dithizone presumably mobilizes thallium from the compartments with maximal concentrations, thus increasing the toxic load of the nervous system (Cavanagh et al., 1974; Ghezzi & Bozza Marrubini, 1979).

Other agents, D-penicillamine and the chelating diethyldithiocarbamate ("dithiocarb"), have also been used as antidotes (Sunderman, 1967; Montoya Cabrera et al., 1979). Dithiocarb caused a three-fold increase in urinary elimination during therapy of a woman (Sunderman, 1967). D-penicillamine was used to treat a patient who initially had 1200 μg thallium/litre in her urine, as well as two other people with thallium poisoning. The authors emphasized that no adverse effects occurred (Alarcón-Segovia et al., 1989). In a detailed comparative survey by Cavanagh et al. (1974), it was stated that for neither of these two antidotes (nor for several other antidotes) was formal proof of benefit available. Negative effects of dithiocarb therapy, such as deterioration of cerebral function, have been observed in patients (Kamerbeek et al., 1971a).

Haemoperfusion does not affect the course of thallium intoxication, according to Heath et al. (1983). Successful treatment by haemodialysis was reported by Barckow & Jenss (1976) and Piazolo et al. (1971). Elimination of thallium by haemoperfusion or haemofiltration should be restricted to intoxications with high doses of thallium during the previous 24 h (Briese & Nessler, 1985b).

A very effective oral antidote in experimental animals and humans is Prussian Blue, potassium ferric hexacyanoferrate(II), an inorganic pigment which is not absorbed by the gut (Heydlauf, 1969; Dvorák, 1970; Kamerbeek et al., 1971b; Günther, 1971; Barbier, 1974; Ghezzi & Bozza Marrubini, 1979; Lehmann & Favari, 1984, 1985). Potassium ions in the molecule are exchanged for thallium ions. Thus, absorption in the intestine is prevented and the thallium-loaded molecule is eliminated with the faeces (Forth & Henning, 1979). This therapy results in faecal elimination greatly exceeding urinary elimination (Stevens et al., 1974). Prussian Blue is now the main therapeutic agent (Forth & Henning, 1979; Lehmann & Favari, 1984; Kazantzis, 1986; Pai, 1987; Chandler et al., 1990), the colloidally soluble form being preferable (de Groot & van Heijst, 1988).

Prussian Blue therapy and forced diuresis with furosemide and mannitol (10 g of the soluble form dissolved in 100 ml 1.5% mannitol as a laxative, twice daily orally or intraduodenally, until urinary thallium elimination is < 0.6 mg/24 h), perhaps supplemented by haemodialysis, is currently considered the optimal therapy for thallium intoxication (Barckow & Jenss, 1976; Forth & Henning, 1979; Briese & Nessler, 1985b; Chandler & Scott, 1986; Wainwright et al., 1988; IPCS, 1992; Aderjan et al., 1994). If Prussian Blue is not available, activated charcoal can be used (IPCS, 1992). The effects on target organs, for instance neurotoxic effects, must be treated symptomatically (Forth & Henning, 1979; Kemper, 1979; Briese & Nessler, 1985b). In laboratory experiments on rats the hepatotoxicity of thallium was prevented by treatment with silymarin, which has been shown to have a hepatoprotective effect in man against several toxic substances (Mourelle et al., 1988).

8.7 Protective measures against excessive occupational exposure

The high toxic potency of thallium has been considered in its TLV or MAK value (threshold limit value or maximum concentration at the workplace) of 0.1 mg/m^3 (Schaller et al., 1980; MT, 1983; Marcus, 1985; DFG, 1990). This value is the limit for a 40-h working week in the USA, France, Germany, United Kingdom and other western countries, and has been reduced in the former-USSR to 0.01 mg/m^3 air (Sabbioni & Manzo, 1980; Nessler, 1985b). The MAK value is the mean value during the normal 8-h working day, and, during this period, only once may a ten-fold higher concentration occur for a period of 30 min (DFG, 1990). According to the West German General Administration Regulation on Air Pollution Control, thallium concentration in dust fall-out should not exceed 0.01 mg/m^2 per day (Ohnesorge, 1985; Ewers, 1988).

On the basis of several reports of recommendations for the protection of employees in industrial plants using thallium, e.g., by Hill & Murphy (1959), Malcolm (1979), Glomme (1983), Nessler (1985b) and a very detailed one by Sessions & Goren (1947), the following protective measures are advisable.

a) General recommendations

i) Access to rooms in which thallium is used should be restricted to a limited number of employees.

ii) Employees should repeatedly be informed about risk and industrial hygiene, in a similar way to employees working with radioisotopes. They should be instructed to report any unusual health symptoms.

iii) Employees should be encouraged to eat potassium-rich food.

b) Medical control

i) By means of a preplacement examination, people suffering from renal, hepatic or neurological diseases, anaemia, blood dyscrasias, hypertension, alcoholism, chronic infections or endocrine gland dysfunction should be excluded from working with thallium.

ii) Urinary thallium should be periodically determined as a means of showing the effects of education programmes and improving industrial hygiene. The intervals will depend on the degree of exposure.

iii) Periodic examinations should pay particular attention to the early toxic effects of thallium, e.g., renal function, gastrointestinal disturbances, the presence of paraesthesia and alopecia.

c) Engineering control

i) Dust scattering should be avoided and handling of thallium should be conducted under exhaust ventilation.

ii) Floors and tables should be wet-mopped.

iii) Dust samplers should be installed for environmental monitoring to permit the evaluation of possible sources of contamination.

d) Personal protective equipment and hygienic measures

i) Employees should be required to use protective work clothes including gloves.

ii) When indicated, personal exposure monitoring should be performed.

iii) Complete sets of personal work clothes should be kept in accommodation separate from that employed for street clothes. Before changing clothes, gloves should be thoroughly washed and then hands, using separate towels.

iv) Depending on the level of exposure, work clothing should be washed periodically.

v) Clothes should be changed before eating, drinking and smoking, all of which should be prohibited at the workplace.

vi) Washing and shower facilities should be provided and their use enforced.

vii) Individual respirators should be worn in all operations producing dust or fumes.

9. EFFECTS ON OTHER ORGANISMS IN THE LABORATORY AND FIELD

Since the majority of results have been obtained in laboratory experiments, field observations on plants (mainly near the cement plant in Lengerich) have not been considered under a special subheading but have been included in sections 4.1.2.2, 9.3.1.2 and 9.3.1.6, and those on vertebrates have been included in sections 9.3.2 and 9.3.3.

The toxicity of thallium has been considered to be comparable with that of mercury and lead, which are its neighbours in the Periodic Table (Emsley, 1978).

9.1 Microorganisms

Buschke & Jacobsohn (1922) observed a sterilizing effect of metallic thallium on various bacteria placed on agar plates. The antibiotic effect of thallium was also recognized during postmortem examination of thallium-poisoned humans, which were less decomposed than other corpses (Muller, 1961).

The most important effect of thallium on microorganisms seems to be the complete or partial inhibition of nitrate formation by *Nitrobacter agilis*, observed at concentrations of 0.8 to 16 mg thallium/litre (Tandon & Mishra, 1969). However, nitrification in soils was only reduced at high thallium concentrations, which also affected plants considerably (McCool, 1933). Direct effects on soil microflora were demonstrated by Drucker et al. (1979). The numbers of total aerobic bacteria and the distribution of other microorganisms were affected at concentrations as low as 1 and 10 mg/kg soil, whereas soil respiration was only reduced at 100 mg/kg soil. Thallium was less toxic than silver, mercury, chromium, cadmium, copper, nickel and zinc (in descending order of toxicity) but more toxic than other elements, e.g., arsenic and lead (Drucker et al., 1979). However, the properties of the soil and the form of thallium used in this study were not reported.

The only known positive effect of thallium on organisms has been described by Richards (1932), who obtained a higher yield of yeast in the presence of 0.1, 1 or 10 mg thallium/litre culture medium. Higher concentrations inhibited the growth of *Saccharomyces cerevisiae*. This positive effect of thallium needs to be verified, since it could have been caused by contamination of the thallium, which would not have been detected by the analytical methods available at that time. Using the same microorganism, Norris et al. (1976) found a growth inhibition of 50% after the addition of 153 mg/litre to liquid culture (which contained 31.3 g potassium/litre) or 10.2 mg/litre to agar medium. The same inhibitory effect with *Escherichia coli* was obtained after the addition of 184.0 mg/litre liquid culture or 10.2 mg/litre agar and with *Bacillus megaterium* after adding 3.1 or 5.1 mg/litre, respectively. The concentrations of thallium (added to the agar medium), which prevented colony formation, i.e. 15.3, 20.4 and 40.8 mg/litre for *B. megaterium*, *S. cerevisiae* and *E. coli*, respectively, also show interspecies differences. Constant thallium concentrations of about 30% of those concentrations which inhibited growth rates by 50%, but with lower concentrations of potassium in the liquid medium, caused a decrease in the growth rates of *B. megaterium* and *S. cerevisiae* (Norris et al., 1976).

Potassium/thallium antagonism has been observed using *Thiobacillus ferrooxidans*. Iron oxidation by growing cultures in a potassium-free medium, but with 0.204 to 204.39 mg thallium (as sulfate) per litre, was normal at 20.4 mg thallium/litre but inhibited by 204.39 mg/litre. However, this inhibitory concentration did not affect iron oxidation in the normal medium containing about 180 mg potassium/litre (Tuovinen & Kelly, 1974).

In inhibitory tests with *Azotobacter chroococcum* and *A. vinelandii* on agar plates, 20 mg thallium sulfate produced very strong inhibition, comparable to that of gold or chromate, and stronger than that of zinc or copper (den Dooren de Jong & Roman, 1971). The two latter heavy metals were found to be also less inhibitory in tests using the bacterium *Klebsiella pneumoniae* (Sikka et al., 1987). However, in a previous study this bacterium was much more sensitive to sulfates of zinc, copper and cadmium than to that of thallium (Wilson & Dean, 1977). The different toxicity mechanisms were also shown by strains resistant to these heavy metals. Thallium-resistant strains showed the opposite reaction and were more sensitive to gentamycin and chloramphenicol than the wild strain. In liquid medium, up to 0.2 mg thallium/litre reduced the doubling time (correlated to the concentration) but induced no lag, whereas higher concentrations also induced a lag of initial development. On agar plates, concentrations of up to 0.3 g/litre were not toxic, but with 0.4 g/litre bacterial colonies developed only rarely. A reduction in the potassium concentrations increased the toxicity of thallium in both media. A synergistic effect, and not a purely additive effect, with zinc and cadmium occurred only at concentrations above 0.1 g/litre (Wilson & Dean, 1977).

The enhanced isolation of mycoplasms in the presence of considerable numbers of other bacteria from the human urogenital tract (Tully et al., 1983) might be caused by a species-specific differences in the sensitivity to thallium, which had been added to the culture medium in addition to penicillin. Such a selective cultivation of bacteria through the addition of different concentrations of thallium to the agar can be used as a taxonomic criterion and should be considered when using clinical material (Kunze 1972a,b).

Different toxicities of thallium(I), thallium(III) and organothallium compounds were shown in a study on bacteria and fungi by Srivastava et al. (1973). At low concentrations (1 and 5 mg/litre), the growth of *Colletotrichum falcatum*, the causative fungus of a sugarcane disease, was affected more by thallium(III) chloride than by thallium(I) chloride. However, at higher concentrations the effect of the thallium(III) compound did not increase, and the fungus was affected more by the monovalent thallium. With both compounds the inhibition did not exceed 20% compared to controls, even after the addition of 80 mg/kg. The organothallium chlorides (diphenyl derivatives) were considerably more toxic than the thallium chlorides, causing a reduction of mycelial growth of 50%, even at concentrations of 13 or 16 mg/kg. This inhibition was obtained with < 1 mg/kg for most ditolyl derivatives (Srivastava et al., 1973).

Five fungal species of the genus *Aspergillus* were similarly sensitive to thallium nitrate. Their sensitivity to cadmium and mercury nitrate was similar, but they were less sensitive to the nitrates of nine other metals (Filimonova et al., 1973).

Anaerobic bacteria are more sensitive to thallium(I) than to organothallium compounds (Huber et al., 1978). In several cases the inhibition of bacterial growth was greater at lower concentrations of the organothallium compounds.

9.2 Aquatic organisms

Little is known about the toxicology of thallium in aquatic systems. Wachs (1988) classified thallium, together with lead and zinc, into the toxic class II. Because of a lack of complex formation, its toxicity is not affected by water hardness, copper, etc. (Zitko et al., 1975). Acute and chronic toxicities to freshwater aquatic life are reported to occur at 1400 and 40 μg/litre, respectively. One species of fish is even affected at 20 μg/litre after 2600 h of exposure. Marine aquatic life seems to be affected at 2130 μg/litre, and as in the case of freshwater organisms, it is clear that more sensitive species than those tested might exist (US EPA, 1980).

9.2.1 Plants

Results from studies on algae and higher aquatic plants are given in Table 26. Photosynthesis is affected, as shown by the reduced oxygen evolution of the algal endosymbiont of the ciliate *Paramecium bursaria* (Di Gaudio & Hirshfield, 1976). In comparison to other heavy metals, higher concentrations of thallium were needed to reduce the light-induced oxygen evolution of the freshwater alga *Chlamydomonas reinhardii* (Overnell, 1975); about 2 mg thallium/litre buffer caused a reduction of about 50%. Thallium strongly affects NADP reduction or the dark stage reactions of photosynthesis. However, it seems to inhibit photosystem I only slightly, and this could not be confirmed in a very detailed later investigation with the green alga *Chlorella saccharophila*, which also showed a reduced oxygen evolution (Wystrcil et al., 1987). Measurements of the variable fluorescence indicated a primary action of thallium on electron transfer in photosystem II. At low concentrations of thallium (2 mg/litre), disturbances of the thylakoid membranes could explain the altered variable fluorescence. A change in the colour of the algal suspension to a greyish green and the alterations of the fluorescence intensity indicated effects on the pigments (Wystrcil et al., 1987). This has also been found for the alga *Chlamydomonas reinhardtii*. β-Carotin showed the greatest sensitivity, followed by chlorophyll a and then chlorophyll b (Maier-Reiter et al., 1987). In addition to an increase in the degradation rate of the pigments, the biosynthesis rate was reduced (section 9.3.1.2).

In a comparison of the toxicity of seven heavy metals for the marine diatom *Ditylum brightwellii*, thallium (thallium(I) chloride) possessed the lowest toxicity, which was not affected by the chelating agent EDTA (Canterford & Canterford, 1980). Within 5 days a 50% growth reduction was obtained with 330 to 350 μg total thallium/litre, corresponding to 0.15-0.17 mg free thallium per litre. To obtain total inhibition, concentrations of about 2.2 times higher were needed (Table 26), but the cells showed an

Table 26. Toxicity of thallium[a] to aquatic plants

Species	Parameter	Exposure time	Concentration of thallium[b]			Reference
			LOEL[c]	EC_{50}[d]	TEC[e]	
Algae						
Chlamydomonas reinhardii	oxygen evolution	0.25 h		2		Overnell (1975)
C. reinhardii	concentration of pigments	22 h	> 0.2	20		Maier-Reiter et al. (1987)
Ditylum brightwellii	growth	5 days		0.34	0.73	Canterford & Canterford (1980)
Dunaliella tertiolecta	growth	-	0.08 0.18[g]			Puddu et al. (1988)
D. tertiolecta	growth	-	0.18[g]			Puddu et al. (1985)

Table 26 (contd).

Species	Parameter	Exposure time	Concentration of thallium[b]			Reference
			LOEL[c]	EC_{50}[d]	TEC[e]	
Elodea canadensis	damage	28 days		2.0		Brown & Rattigan (1979)
E. canadensis	oxygen evolution	24 h		1.4		Brown & Rattigan (1979)
Phaeodactylum tricornutum	growth	-	0.14 0.22[g]			Puddu et al. (1988)
P. tricornutum	growth	-	0.24[g]			Puddu et al. (1985)
Selenastrum capricornutum	concentration of pigments	96 h		0.11		US EPA (1978)
S. capricornutum	growth	96 h		0.10		US EPA (1978)

Table 26 (contd).

Species	Parameter	Exposure time	Concentration of thallium[b]			Reference
			LOEL[c]	EC_{50}[d]	TEC[e]	
Higher Plants						
Lemna minor	growth (weight)	10 days	0.014 281.5[h]	0.04 510.0[h]		Kwan & Smith (1988)
L. minor	growth (frond number)	10 days	0.016 293.8[h]	0.047 550.1[h]		Kwan & Smith (1988)
L. minor	growth (frond area)	10 days	0.008 195.8[h]	0.033 443.7[h]		Kwan & Smith (1988)
L. minor	damage	28 days		0.008		Brown & Rattigan (1979)

[a] Thallium(I), unless otherwise stated
[b] mg thallium/litre nutrition solution, unless otherwise stated
[c] LOEL = low-observed-effect level
[d] EC_{50} = concentration at which the life parameters were reduced by 50%
[e] TEC = concentration at which life parameters were totally inhibited
[f] - = no data given
[g] Thallium(III)
[h] mg thallium/kg dry weight of plant

abnormal appearance, i.e. pseudo-resting spore-like cells and shrinkage of protoplast and concentration of chromophores, within about 3 days (Canterford & Canterford, 1980; Canterford, 1980).

Species-specific sensitivities, but also different effects of the monovalent and the trivalent forms of thallium, were evident in two marine algae, *Phaeodactylum tricornutum* and *Dunaliella tertiolecta* (Table 26) (Puddu et al., 1985, 1988). Complex formation with EDTA reduced the toxicity of the thallium(III) compound. Using different strains of three acidophilic algae, the type and level of inhibition of growth after culture in a medium containing thallium sulfate was very similar in the strains of *Cyanidium caldarium* and *Cyanidioschyzon merolae*, whereas strains of *Galdieria sulphuraria* showed different effects (Albertano & Pinto, 1986).

Generally, freshwater algae are affected at concentrations as low as 100 μg/litre (US EPA, 1980). In the submerged macrophyte *Elodea canadensis*, 1.4 and 2.8 mg thallium/litre water (as thallium(I) sulfate) reduced the photosynthetic oxygen evolution during 24 h by 50% and 90%, respectively (Brown & Rattigan, 1979). In parallel experiments, the free-floating *Lemna minor* was much more sensitive (Table 26). The uptake and accumulation of thallium (thallium(I) acetate) and its effects on growth parameters of *L. minor* (frond area, fresh weight, frond number) were determined by Kwan & Smith (1988). High concentrations of 1 and 2 mg/litre water induced chlorosis by the eighth day, and 2 days later the fronds were completely devoid of any colour. Comparing the number of fronds at different concentrations of thallium during a period of 10 days, a stimulatory effect was evident after exposure to 2 μg/litre and 10 μg/litre at the end of the period. However, smaller fronds were produced and, therefore, the surface area covered by *Lemna* also decreased at the lowest thallium concentrations. At 20 μg/litre or more, all growth parameters were reduced. Subsequent culture in thallium-free water resulted in good recovery from the previous thallium exposure, provided that the concentration had not exceeded 20 μg/litre. The bioconcentration factor (based on plant weight) was 88 000 at the lowest exposure (2 μg/litre) and 6000 at an exposure of 153 μg/litre.

Using 40 μg/litre growth medium, the thallium concentrations in *L. minor* did not further increase after 140 h, and over 80% of the thallium remained in the plant (Smith & Kwan, 1989). In plant homogenates, thallium showed little association with proteins, and the reactions of the thallium in the soluble fraction were comparable to those of the free metal ion. Like potassium, thallium accumulates in the cell vacuole. When various concentrations of thallium were added to the medium (0.02 to 0.2 mg/litre), about 0.04 mg/litre caused a 50% reduction of growth after a period of 10 days (Smith & Kwan, 1989). Since concentrations of 0.05 to 0.09 mg/litre were found in rivers contaminated by mining (Zitko et al., 1975), significant effects on macrophytes in such rivers would be expected, given the long exposure period.

9.2.2 Animals

Toxicity data on aquatic animals are summarized in Table 27. In an immobilization test on *Daphnia magna*, it was shown that the toxicity of thallium(I) nitrate was higher than that of nickel or cadmium but lower than that of copper, silver or mercury (Bringmann & Kühn, 1982). In general, arthropods were affected at lower concentrations than fish. The 96-h LC_{50} values were much lower for the daphnids (water fleas) *Daphnia magna* and *Mysidopsis bahia* (2.2 mg/litre) than for the freshwater fish *Lepomis macrochirus* (bluegill) (120 mg/litre), although 50% of fathead minnows (*Pimephales promelas*) were killed by 0.86 mg/litre (LeBlanc, 1984).

Comparing vertebrate with invertebrate marine species, about 10% of the concentration needed to kill 50% of sheepshead minnows or tidewater silversides was sufficient to kill 50% of marine shrimps (US EPA, 1980). Acute values were 3- to 32-fold higher than chronic values (US EPA, 1980).

Considerable differences between species were also evident in a study by Nehring (1962/63). This study also showed that in perch (*Perca fluviatilis*) lethal thallium effects depend on the length of exposure. The fish were killed by 200 or 500 mg/litre within 10 h, while concentrations of 100, 90-50, 40 and 20 mg/litre caused death within 2, 2.5-3.5, 7.5 and 14 days, respectively. At 15 mg/litre, the perch survived at least 17 days. Trout (*Salmo gairdneri*) and roach (*Rutilus rutilus*) were more sensitive and died within 8 and 14 days, respectively, at 4 mg/litre. In these species no effects were observed within 17 days at concentrations of 2 mg/litre.

Juvenile Atlantic salmon (*Salmo salar*) are exceptionally sensitive to thallium contamination (Zitko et al., 1975). The

Table 27. Toxicity of thallium[a] to aquatic animals

Species	Stage	Parameter	Exposure time	Concentration of thallium			Reference
				LOEL[b]	EC_{50}[c]	TEC[d]	
Invertebrates							
Daphnids	-[e]	survival	48 h		2.2		LeBlanc (1984)
Daphnia magna	juvenile	mobility	24 h	> 0.003	0.11	0.95	Bringmann & Kühn (1982)
Daphnia sp.	adult	survival	72 h	2-4			Nehring (1962/63)
Gammarus sp.	adult	survival	72 h	4			Nehring (1962/63)
Mysid shrimp (*Mysidopsis bahia*)		survival	96 h		2.1		US EPA (1978)
Vertebrates							
Atlantic salmon (*Salmo salar*)	juvenile	survival	47 h		10		Zitko et al. (1975)
			112 h		1		
			435 h		0.1		
			2600 h	0.03			

Table 27 (contd).

Sheepshead minnow (*Cyprinodon variegatus*)	adult	survival	96 h		20.9		US EPA (1978)
Bluegill (*Lepomis macrochirus*)	adult	survival	96 h		132		US EPA (1980)
	adult	survival	96 h		120		LeBlanc (1984)
Perch (*Perca fluviatilis*)	adult	survival	72 h	60			Nehring (1962/63)
Fathead minnow (*Pimephales promelas*)	embryo	hatch		> 0.2		< 0.72	LeBlanc & Dean (1984)
	larva	survival	30 day	< 0.04		< 0.35	
	larva	growth	30 day	> 0.04		> 0.2	
	adult	survival	96 h		0.86		LeBlanc (1984)
	adult	survival	96 h	0.08			US EPA (1980)
	adult	survival	96 h		1.8		US EPA (1980)
Roach (*Rutilus rutilus*)	adult	survival	72 h	40-60			Nehring (1962/63)

Table 27 (contd).

Species	Stage	Parameter	Exposure time	Concentration of thallium			Reference
				LOEL[b]	EC_{50}[c]	TEC[d]	
Rainbow trout (*Salmo gairdneri*)	adult	survival	72 h	10-15			Nehring (1962/63)
Tidewater silverside (*Mendia berrylina*)		survival	96 h		24		Dawson et al. (1975/77)
Toad	adult	survival	-			16.7[g]	Swain & Bateman (1909/10)

[a] mg thallium/litre water
[b] LOEL = low-observed-effect level
[c] EC_{50} = concentration at which the life parameters were reduced by 50%
[d] TEC = concentration at which life parameters were totally inhibited
[e] - = data not given
[f] 12 days in experiments with other species
[g] mg/kg injected into a lymph sinus

18-day LC_{50} is 0.1 mg/litre. The mortality of fish exposed to 0.03 mg/litre was reported to be equal to that of the controls. The authors suggested that 0.02 mg/litre be regarded as a no-observed-effect level (NOEL), and that 0.04 mg/litre be regarded as a low-observed-effect level (LOEL). Mixtures with copper or zinc did not alter thallium toxicity.

Behavioural alterations in perch were reported by Nehring (1962/63), even at low concentrations. Initially, following exposure to thallium, food consumption increased, but after one or two days neuronal damage occurred. This was indicated by uncoordinated movement, paralysis of gills and disturbance of balance. Similar effects on movement and respiration in fish had already been observed by Swain & Bateman (1909/10).

Larvae of fathead minnows (*Pimephales promelas*) were much more sensitive than the embryos to thallium sulfate (LeBlanc & Dean, 1984). No effect on the rate of hatching was observed at concentrations up to 200 μg/litre, but at 720 μg/litre no hatching occurred. At 350 μg/litre no larvae survived. Growth was affected at 120 μg/litre.

In a study by Birge (1978), rainbow trout (*Salmo gairdneri*) eggs were exposed to thallium from fertilization to 4 days after hatching (total of 28 days). The exposure water was renewed every 12 h. An LC_{50} of 0.17 mg/litre and an LC_1 of 0.0084 mg/litre were reported. Goldfish (*Carassius auratus*) eggs were also exposed to thallium from fertilization to 4 days after hatching (total of 7 days). The exposure water was renewed every 12 h. An LC_{50} of 7.00 and an LC_1 of 0.0525 mg/litre were reported (Birge, 1978).

Amphibia are also affected by thallium. Development of frog spawn was unaffected by concentrations of 40.8 and 200 mg/litre, but a concentration of 0.4 mg/litre killed all tadpoles on hatching (Dilling & Healey, 1926). This indicates that the absorption of thallium by the eggs was minimal. Injections of lethal concentrations of thallium acetate (> 0.005 g) into the lymph sinus of adult toads (Table 27) caused loss of control of the hindlimbs and death by asphyxia (Swain & Bateman, 1909/10). In a study on the narrow-mouth toad (*Gastrophryne carolinensis*), eggs were exposed to thallium from fertilization to 4 days post-hatch (total 7 days). The exposure water was changed every 12 h. An LC_{50} of 0.11 and an LC_1 of 0.0024 mg/litre were reported (Birge, 1978).

9.3 Terrestrial organisms

9.3.1 Plants

Early investigations into the toxicity of thallium to plants were summarized by Scharrer (1955). The most obvious effects are decreased productivity, inhibition of photosynthesis and direct cytotoxicity. Toxicity data are listed in Table 28.

9.3.1.1 Plant photosynthesis

In a study by Bazzaz et al. (1974), the net photosynthesis of excised tops of sunflowers (*Helianthus annuus*) decreased both with time and with the concentration of thallium in the nutrient solution (2 to 200 mg/litre). At the highest concentration, photosynthesis was inhibited by about 70% after 1 day, and the plants began to wilt. At the lower concentrations, these visible symptoms appeared 4 days later. There was a strong log-linear relationship between photosynthesis and the thallium content of the plants. Stomatal opening was reduced by 30 and 90%, respectively, at concentrations of 0.2 and 2 mg thallium/litre, but increased concentrations only caused a slight additional effect (Bazzaz et al., 1974). In a direct comparison Carlson et al. (1975) observed that the effects of thallium sulfate on photosynthesis and transpiration in sunflowers were similar to those in maize (*Zea mays*) at low thallium concentrations (up to 2 mg thallium/litre solution). Higher concentrations, up to 10 mg/litre, induced further inhibition in maize but not in sunflowers. Using the data of Carlson et al. (1975), there was a good linear correlation for maize between the occurrence of stomatal opening (y) and thallium content of the solution (x):

$$y = 66.4 - 0.4\,x;\ \text{regression coefficient} = 0.93$$

This was greater than the correlation calculated using the data for sunflowers:

$$y = 45.1 - 0.4\,x;\ \text{regression coefficient} = 0.66.$$

The effects of thallium on the photosynthesis of spinach chloroplasts have been investigated by Wystrcil et al. (1987). Some alterations of variable fluorescence indicated a primary action of thallium on electron transfer in photosystem II, which was also evident in green algae (section 9.2.1). In photosystem I, superoxide dismutase might also be affected by thallium (Wystrcil et al., 1987).

Table 28. Toxicity of thallium[a] to terrestrial plants[b]

Species	Parameter	Exposure time	Concentration of thallium[c]			Reference
			LOEL[d]	EC_{50}[e]	TEC[f]	
Brassica napus	shoot growth	10 days	> 20[h,n]			Makridis & Amberger (1989b)
B. napus	shoot growth	18 days	800[i,n] < 2[h,n]			Makridis & Amberger (1989b)
B. napus	shoot growth	28 days	> 2669 > 10.0[h]			Allus et al. (1987)
B. napus	root growth	28 days	> 760 > 10.0[h]			Allus et al. (1987)
Cucumis sativus	growth	10 days	< 10[h]			Puerner & Siegel (1972)
Garden lettuce	growth	7 days	> 100 10[h]			Schweiger & Hoffmann (1983)
Garden lettuce	growth	summer	30 < 10[k]		> 500[k]	Hoffmann et al. (1982)
Green kale	growth	7 days	> 500 10[h]			Schweiger & Hoffmann (1983)

Table 28 (contd).

Species	Parameter	Exposure time	Concentration of thallium[c]			Reference
			LOEL[d]	EC_{50}[e]	TEC[f]	
Helianthus annuus[g] (sunflower)	photosynthesis	4-5 days		63		Bazzaz et al. (1974)
H. annuus	photosynthesis	4-9 days		82		Carlson et al. (1975)
H. annuus	growth	7 days	> 100 10[h]			Schweiger & Hoffmann (1983)
H. annuus[g]	stomata opening	8 h	< 0.2[h]	~0.8[h]		Bazzaz et al. (1974)
Hordeum vulgare	shoot growth		20 0.5[h]			Davis et al. (1978)
H. vulgare	shoot growth	28 days	> 21 < 0.2[h]			Allus et al. (1987)
H. vulgare	root growth	28 days		< 86 < 0.2[h]		Allus et al. (1987)
Kohlrabi (young)	growth	summer	> 600 500[k]			Hoffmann et al. (1982)

Table 28 (contd).

Lolium perenne	shoot growth	21 days	0.71		251.2	Al-Attar et al. (1988)
L. perenne	root growth	21 days	2.1		1990	Al-Attar et al. (1988)
Nicotiana tabacum[p]	survival	24 h		0.02[h]		Siegel (1977)
N. tabacum	germination	24 h		0.02[h]		Siegel (1977)
Phaseolus vulgaris	growth	33 days		< 0.55[h]		Kaplan et al. (1990)
P. vulgaris	shoot growth	10 days	130[i,n] < 1[h,n]			Makridis & Amberger (1989b)
Pisum sativum	stem growth	28 days	5-10 210-360[h] 5-10[n] 115-123[h,n]			Pötsch & Austenfeld (1985) Pieper & Austenfeld (1985)
Pisum sativum	leaf growth	28 days	1-5 30-75[h] > 10[n] 30-43[h,n]			Pötsch & Austenfeld (1985) Pieper & Austenfeld (1985)

Table 28 (contd).

Species	Parameter	Exposure time	Concentration of thallium[c]			Reference
			LOEL[d]	EC_{50}[e]	TEC[f]	
P. sativum	root growth	28 days	> 10 > 180[h]			Pötsch & Austenfeld (1985)
			> 10[n] > 80[h,n]			Pieper & Austenfeld (1985)
Radish[l]	growth	summer	35 < 500[k]			Hoffmann et al. (1982)
Radish[m]	growth	summer	< 300 > 500[k]			
Rape	growth	summer	> 500 200[k]		500[k]	Hoffmann et al. (1982)
Spinach	concentration of pigments	14 days		< 150 > 0.2[h]		Maier-Reiter et al. (1987)
Spinach	growth	9 days	280[i] 2[h,i]			Schweiger & Hoffmann (1983)

Table 28 (contd).

Vicia faba	stem growth	28 days	> 10 > 222[h] 5-10[n] 36-76[h,n]	Pötsch & Austenfeld (1985) Pieper & Austenfeld (1985)
V. faba	leaf growth	28 days	> 10 > 8[h] 5-10[n] 5-7[h,n]	Pötsch & Austenfeld (1985) Pieper & Austenfeld (1985)
V. faba	root growth	28 days	> 10 > 1320[h] > 10[n] > 575[h,n]	Pötsch & Austenfeld (1985) Pieper & Austenfeld (1985)
Zea mays (corn)	growth	7 days	> 100 10[h]	Schweiger & Hoffmann (1983)
Z. mays	root growth	28 days	1[h]	Logan et al. (1984)
Z. mays	shoot growth	28 days	1[h]	Logan et al. (1984)
Z. mays	root growth	28 days	1[h,n]	Logan et al. (1984)
Z. mays	shoot growth	28 days	1[h,n]	Logan et al. (1984)

Table 28 (contd).

Species	Parameter	Exposure time	Concentration of thallium[c]			Reference
			LOEL[d]	EC_{50}[e]	TEC[f]	
Z. mays	photosynthesis	4-9 days		82		Carlson et al. (1975)
Z. mays[o]	stomata opening	8 h		2[h]		Carlson et al. (1975)

[a] Thallium(I), unless otherwise stated
[b] Whole plants, unless otherwise stated
[c] mg thallium/kg dry weight of plant tissue, unless otherwise stated
[d] LOEL = low-observed-effect level
[e] EC_{50} = concentration at which the life parameters were reduced by 50%
[f] TEC = concentration at which life parameters were totally inhibited
[g] Tops
[h] mg thallium/litre nutrition solution
[i] Concentration at which the life parameters were reduced by 10%
[k] mg thallium/kg soil
[l] Root
[m] Leaf
[n] Thallium(III)
[o] Epiderm
[p] Protoplast
[q] Seed

9.3.1.2 Cytotoxic effects

Chlorosis, followed by marginal necrosis of the leaves, is the most prominent sign of thallium toxicity in plants. Different courses of thallium poisoning in various plant species were reported by McCool (1933) and later by Spencer (1937) in tobacco, by Carlson et al. (1975) in corn and sunflowers, by Davis et al. (1978) in barley, by Makridis & Amberger (1989b) in bushbeans and rape, and by Kaplan et al. (1990) in beans. Similar observations on the leaves of trees in the vicinity of the cement plant in Lengerich, Germany were reported by LIS (1980).

The course and location of chlorosis seemingly depend on thallium concentrations in the substrate and presumably correspond to the distribution of thallium in the plant (Schweiger & Hoffmann, 1983).

In isolated protoplasts of *Nicotiana tabacum*, a cytotoxic effect was also observed; 10% and 50% had lysed after a 24-h incubation in 4 (± 0.4) and 20 (± 2) μg thallium/litre, respectively, irrespective of the age of the protoplasts (Siegel, 1977). These values are nearly identical to the percentages of seed in which germination was inhibited (Siegel, 1977).

Chlorosis indicates a reduced concentration of pigments (section 9.2.1). In spinach, it is firstly the concentration of β-carotene which is reduced, then that of chlorophyll *a* and finally that of chlorophyll *b*. The concentrations of β-carotene and chlorophyll *a* were about half the normal value after 2 weeks of incubation in a hydroculture medium containing thallium nitrate (0.2 mg thallium/litre) (Maier-Reiter et al., 1987).

9.3.1.3 Growth of plants

Adverse effects of thallium on the growth of plants have been reported for various test systems (Table 28). In a comparison of the effects of three heavy metals on the growth of cabbage seedlings, cadmium and thallium were found to be less toxic than silver (Allus et al., 1988).

Initial mycelial growth of three fungal species was inhibited on agar plates containing 0.25 or 0.50 mg thallium/litre (Seeger & Gross, 1981).

In tobacco plants, concentrations as low as 5 mg thallium(I) per litre inhibited terminal growth and caused a temporary outgrowth of axillary buds all resembling natural frenching, i.e. a reticulate interveinal chlorosis. In hydrocultures the root system was strongly affected after 12 days at 0.067 mg thallium/litre. Thallium(I) nitrate and sulfate were similarly toxic. The toxic effects of lower concentrations were reduced by the addition of aluminium sulfate, nitrogen and potassium iodide. In other species of *Nicotiana*, only terminal growth, chlorophyll formation or roots were affected, and the level of sensitivity to thallium corresponded to the level of susceptibility to frenching in the field (Spencer, 1937).

After being watered for 15 days with 20 or 200 mg thallium per litre, the growth of cucumber seedlings was unaffected, but growth was reduced by 2000 mg thallium/litre. Toxicity was increased by limiting the uptake of potassium. The higher sensitivity of the epicotyls compared to the hypocotyls indicated that cell multiplication processes are more sensitive than those entailing cell enlargement and differentiation, a phenomenon known from many other stress factors and toxic substances (Siegel & Siegel, 1976). In corn, production of top and root biomass was severely reduced to between 50 and 60% of the controls (Carlson et al., 1975).

From the differential reduction in weight in parts of the bean plant, it has been shown to be possible to rank them according to their increasing sensitivity to thallium(I): roots >> upper leaves > lower leaves = upper stems > lower stems. Results from hydroponic cultures were similar to those from field studies (Kaplan et al., 1990). In another variety of bean, the weights of leaves and stems, but not those of roots, were affected by exposure to thallium(III) (up to 2 mg/kg). However, thallium(I) had no effect (Pötsch & Austenfeld, 1985; Pieper & Austenfeld, 1985).

Garden lettuce and radish growing in soils treated with $TlNO_3$ exhibited considerably reduced growth at concentrations in dry plant tissue of 30-35 mg thallium/kg (Hoffmann et al., 1982). Growth of perennial rye grass (*Lolium perenne*) was adversely affected when concentrations of thallium exceeded about 0.7 mg/kg dry weight in shoots and 2.0 mg/kg dry weight in roots (Al-Attar et al., 1988).

3.1.4 Different sensitivities to thallium(I) and thallium(III)

Only small differences were observed between the toxic effects of thallium(I) and thallium(III) on the dry weight of roots and shoots of maize. Growth was slightly more reduced after application of thallium(I) (Logan et al., 1984). Similarly, in two detailed studies of the effects of thallium(I) and thallium(III) nitrate (0.2, 1, 2 mg thallium/litre nutrient solution) on the dry weight of pea plants, growth was found to be affected more after exposure to thallium(I). However, completely opposite results were obtained for field beans (Pötsch & Austenfeld, 1985; Pieper & Austenfeld, 1985). Following complexation of thallium(III) and thallium(I) nitrate with EDTA, plants reacted differently to the two compounds. In bean stems and roots, but not in leaves, concentrations of thallium were increased by thallium(III) EDTA compared to those resulting from thallium(III) on its own, while thallium(I) EDTA resulted in similar or lower thallium concentrations than thallium(I) on its own in all three plant organs. In the stems and leaves of peas, thallium(III) EDTA resulted in lower thallium concentrations than thallium(III) on its own, while in the roots thallium levels were the same for both salts. Consistently higher thallium concentrations were found in leaves, stems and roots of peas after exposure to thallium(I) EDTA, compared to thallium(I) on its own (Pötsch & Austenfeld, 1985; Pieper & Austenfeld, 1985).

3.1.5 Concentration of trace elements

The effects of thallium on plants could be caused mainly by an imbalance of essential cellular monovalent cations or by a disturbed uptake of trace elements (Yopp et al., 1974; Schweiger & Hoffmann, 1983). As can be concluded from the data summarized in Table 29, thallium seems to have no uniform effect on the trace element content; the differences between the two investigations using beans are striking. In most studies the concentration of magnesium was found to be reduced.

Exposure to thallium(III) chloride reduced the concentrations of potassium and trace elements such as copper, zinc and iron in bush beans by up to 20%; calcium, magnesium and manganese were only slightly affected (Makridis & Amberger, 1989b). Rape was less sensitive; uptake was decreased for potassium and copper, but increased for zinc (and for calcium, magnesium, manganese and iron by the reduced growth).

Table 29. Thallium-induced changes in uptake[a] or concentrations of trace elements in plants

		Part of plant	Trace elements								Reference
			B	Ca	Cu	Fe	Mg	Mn	Mo	Zn	
Thallium(I)											
Compound	$TlNO_3$	shoot	-	-	d	-	-	i	-	d	Schweiger & Hoffmann (1983)
Concentration[b]	10										
Exposure time	7 days										
Species	sunflower										
Compound	Tl_2SO_4	root	u	u	i	u	d	i	u	i	Kaplan et al. (1990)
Concentration	1	leaf	i[c]	d	u	u	d	u	i[c]	i[c]	
Exposure time	33 days										
Species	bean										
Compound	$TlNO_3$	root	-	-	u	u	-	d	-	d	Pötsch & Austenfeld (1985)
Concentration	0.2-2	stem	-	-	u	u	-	d	-	u	
Exposure time	28 days	leaf	-	-	u	u	-	u	-	u	
Compound	+EDTA	root	-	-	u	u	-	d	-	u	
Species	bean	stem	-	-	u	u	-	u	-	u	
		leaf	-	-	u	u	-	u	-	u	

Table 29 (contd).

Compound	$TlNO_3$	root	-	-	d	u	-	d	-	d	Pötsch & Austenfeld (1985)
Concentration	0.2-2	stem	-	-	u	i	-	d	-	u	
Exposure time	28 days	leaf	-	-	u	u	-	u	-	u	
Compound	+EDTA	root	-	-	u	u	-	d	-	u	
Species	pea	stem	-	-	u	i	-	i	-	u	
		leaf	-	-	i	i	-	u	-	i	
Thallium(III)											
Compound	$Tl(NO_3)_3$	root	-	-	u	u	-	d	-	d	Pieper & Austenfeld (1985)
Concentration	0.2-2	stem	-	-	u	u	-	d	-	i	
Exposure time	28 days	leaf	-	-	u	u	-	d	-	u	
Compound	+EDTA	root	-	-	u	u	-	d	-	u	
Species	pea	stem	-	-	u	i	-	i	-	u	
		leaf	-	-	i	i	-	i	-	i	
Compound	$Tl(NO_3)_3$	root	-	-	u	u	-	d	-	d	Pieper & Austenfeld (1985)
Concentration	0.2-2	stem	-	-	u	u	-	d	-	u	
Exposure time	28 days	leaf	-	-	u	u	-	d	-	d	
Compound	+EDTA	root	-	-	u	u	-	d	-	u	
Species	bean	stem	-	-	u	u	-	u	-	u	
		leaf	-	-	u	u	-	u	-	u	

Table 29 (contd).

		Part of plant	Trace elements								Reference
			B	Ca	Cu	Fe	Mg	Mn	Mo	Zn	
Compound Concentration Exposure time Species	$TlCl_3$ 10-20 10 days bean		-	i	d	d	u	i	-	d	Makridis & Amberger (1989b)
Compound Concentration Exposure time Species	$TlCl_3$ 10-20 10 days rape		-	i[d]	d	i[d]	i[d]	i[d]	-	i	Makridis & Amberger (1989b)

[a] d = decrease; u = unchanged; i = increase; - = not determined
[b] mg/litre solution
[c] Only upper not lower leaves
[d] Increase by reduced growth

Complex effects on trace elements were found in studies with *Pisum sativum* and *Vicia faba* in which thallium(I) and thallium(III) nitrate and their respective EDTA complexes were used (Table 29) (Pötsch & Austenfeld, 1985; Pieper & Austenfeld, 1985).

9.3.1.6 Sensitivity of plants

Differing sensitivities among plant species, strains and individuals have been reported for a number of air-borne contaminants (Guderian, 1977). Species differences were also evident in many investigations of thallium, including the detailed studies carried out at Lengerich, Germany (LIS, 1980). There the residues of pyrite roasting had been used for six years before effects were obvious (Gubernator et al., 1979). Coniferous trees were not affected, oaks only slightly, and summer lime trees far more than winter ones. Sweet cherry trees were more sensitive to thallium than sour cherry trees. The leaves of pear trees still appeared healthy when apple and plum trees had already lost theirs. Very sensitive vegetables included beans, cucumber and potatoes. The sensitivity of fodder plants varied too; the yield of maize was strongly reduced, but not that of rape or turnip. There seems to be a tendency for plants with a "hard" leaf surface to be less affected than those with a soft, hairy surface (LIS, 1980).

Mechanisms of resistance to thallium seem to vary. Comparing the growth data of beans (section 9.3.1.3) and peas, the higher tolerance of beans to thallium(I) and thallium(III) (section 9.3.1.4) corresponds to a higher concentration of thallium in the roots than in the stems, which in turn contained more thallium than the leaves (Pötsch & Austenfeld, 1985; Pieper & Austenfeld, 1985). The more sensitive peas possessed high thallium concentrations in the stems, followed by those in the roots and leaves. Just the opposite distribution is evident when barley and rape are compared: a higher concentration in roots than shoots in the susceptible barley and the opposite in the tolerant rape (Allus et al., 1987).

Green rape possesses a much higher resistance than beans after application with thallium(III) chloride. Roots were more sensitive than shoots and became brown. Using a 2 μg/litre nutrition solution, the growth of bush beans was increasingly reduced after 3 days. The first signs of reduced growth of rape were observed after 8 days incubation in a solution of 5 mg/litre; other symptoms could only be recognized after the thallium concentration was increased to 10 mg/litre. Then growth of the roots decreased and they became brown (Makridis & Amberger 1989b). Thus, two mechanisms seem to enable survival at high thallium concentrations: beans (and perhaps barley, but not peas) reduce the amount of thallium which is transported to the leaves, whereas the leaves of rape are only affected at very high concentrations of thallium. Tolerance to relatively high concentrations of thallium may be a result of complexation in plant tissue, a phenomenon observed with other heavy metals (Cataldo & Wildung, 1978).

In addition, selective pressure can increase tolerance. The morphology of plants in the Alsar region of Yugoslavia, with a naturally high soil thallium concentration, is not affected (Zyka, 1972), whereas such concentrations cause severe damage at other locations (Schoer, 1984). In the Alsar region, zonation of plant species with respect to soil thallium concentration can be observed (Zyka, 1972), indicating the levels which are toxic to the various species.

9.3.2 Wild animals

The effects of thallium on invertebrates have rarely been investigated. After ant workers ingested about 0.2 mg thallium chloride or thallium acetate per insect over a period of 2 months, all survived (Jeantet et al., 1977). Field baits containing thallium sulfate or acetate have been used against the fire ant and Pharaoh's ant and destroyed about 90% of the colonies. These studies and the toxic effects on crickets have been summarized by Negherbon (1959).

Corn poisoned with thallium sulfate has been used on a large scale to control rodents. Such pest control carried out in the field can affect various seed-eating animals and their predators (Munch et al., 1974). An oral LC_{50} of 35 mg/kg fresh weight in starlings (*Sturnus vulgaris*) was reported by Schafer (1972). This LC_{50} was calculated following a single dose of thallium sulfate administered via gavage, with a 7-day observation period. Thallium-poisoned baits have also been used to control predatory birds. It was presumed that 9 out of 37 bald and golden eagles, which were collected sick or dead in the USA, died from thallium poisoning in 1971-1972. Their kidneys contained high concentrations of thallium (14 to 63 mg/kg) (Cromartie et al., 1975). In experimentally poisoned eagles, which died from a single oral dose of 120 mg thallium/kg body weight, the kidneys contained 39 and 104 mg/kg (Bean & Hudson, 1976). Linsdale (1931) reported the toxic effects of excessive use of thallium in California for "ground squirrel control" on 58 species of game birds, song birds and other wild animals. In 1972 all use of thallium in pesticides was banned in the USA (Smith & Carson, 1977); in a study on birds carried out between 1977 and 1981 no elevated thallium levels could be detected (Wiemeyer et al., 1986).

In Denmark, partridges, pheasants, red foxes, badgers and martens were found to be killed by direct ingestion of thallium-containing rodenticides or poisoned prey (summarized by Munch et al., 1974; Clausen & Karlog, 1974). Patho-anatomical findings from 1963 to 1971 indicated that 55 out of 299 red foxes and 5 out of 17 badgers examined had suffered from poisoning. Determinations of thallium concentrations in the kidney, liver and intestines demonstrated that 27 foxes and 1 badger had presumably been killed by thallium (Table 15). In most of the foxes not suspected of thallium poisoning, thallium concentrations were < 0.1 mg/kg. Several of the poisoned foxes showed abnormal behaviour, but only one fox showed clear hair-loss. A typical sign in poisoned foxes was an empty stomach (Munch et al., 1974). This was not observed in poisoned martens and badgers. In addition, there were no skin lesions. Before death, many of the martens showed uncoordinated movements and loss of balance. The thallium concentrations in the inner organs of martens and badgers ranged up to 92 mg/kg wet weight (Table 15) (Clausen & Karlog, 1974).

9.3.3 Household pets and farm animals

Accidental poisoning of pet animals (dogs and cats), ducks and pigeons has been reported repeatedly (Zook & Gilmore, 1967). The first detailed investigations considering toxic effects in dogs and cats were performed to evaluate the risk of rodent poisoning campaigns with thallium sulfate. Such cases used to be numerous, but in recent years only occasional poisonings have occurred, due to the reduced use of thallium as a rodenticide. For example, in Baden-Württemberg, Germany, only 6 dogs, 4 cats and some ducks and pigeons were found to have been poisoned by thallium from 1977 to 1989, and then no case occurred up to 1992 (F. Baum, 1993, Institute of Animal Hygiene, Freiburg, Germany; personal communication to the IPCS).

Symptoms of chronic intoxications in pets and farm animals are similar to those of acute intoxications and can usually best be observed in dogs. In ruminants uncharacteristic symptoms develop (Hapke, 1984).

In some areas with naturally very high thallium levels, e.g., in Yugoslavia and Israel, natural poisoning of farm animals after consumption of vegetation has occurred (summarized by Gough et al., 1979).

Table 30 summarizes early investigations by Ward (1931) and Shaw (1932) on the toxicity of thallium sulfate to farm animals (quails, geese, ducks and cattle), carried out in order to assess the risks of its use as a rodenticide. In ducks an intraperitoneal injection of up to 10 mg thallium/kg did not affect the birds, 15-25 mg/kg caused loss of feathers on the back and 35-100 mg/kg was lethal within 24-63 h (Ward, 1931). After feeding barley contaminated with 35, 50, or 75 mg thallium/kg, ducks survived, or died in 12 days, or overnight, respectively. Those ducks which died showed a mucous clogging of the nasal passages (resulting in a marked gasping for breath), profuse and green-coloured diarrhoea, loss of accommodation, wobbly gait and extreme exhaustion. Death was due to respiratory failure and occurred within 2 h after the beginning of intermittent asphyxial spasms. Dissection of the dead ducks demonstrated that the intestinal tract was plugged with a thick yellowish mucous. In addition, irritations and ulcerations were present in the small intestine, and the livers were enlarged and degenerated (Ward, 1931).

The same author investigated the effect on cattle (Ward, 1936). Using thallium(I) sulfate, one cow received 50 mg thallium/kg body weight and 3 heifers 35, 25 and 15 mg/kg, the latter two additional doses of 20 and 35 mg/kg at 69 and 31 days after the first administration. Two of the animals defaecated small quantities of bloody faeces, all showed muscular twitching of flank and drooling of a stringy mucous from nose and mouth. The cow died 5 days after administration, and the two heifers with the highest doses lived 11 and 14 days. The last animal was killed 3 days after the last administration. Pathological changes were evident in the lymphatic vessels (congested and oedematous), the liver (pale or congested) and kidney (congested) and the walls of the digestive tract (haemorrhages, ulcerations). Other organs appeared normal, and no significant depilatory effect occurred. Recently Frerking et al. (1990) reported thallium poisoning in cattle caused by the use of contaminated silage. Symptoms were muscular twitching, colic, nervous behaviour, extreme thirst, drooling from nose and mouth, loss of hair at the tail and, later, erosion of nasal epithelium. The authors estimated that over a period of 6 weeks the cows had ingested 0.75 mg thallium/kg body weight daily (a total of 17 g thallium).

Table 30. Acute toxicity of thallium(I) sulfate for farm animals

Species	Route of administration	Period of observation	Toxicity[a]	Dose (mg thallium/kg body weight)	Reference
Quail	oral	7 days	LC_{100}	~12	Shaw (1932)
Goose	oral	14 days	LC_{100}	~15	Shaw (1932)
	oral	2-3 days	LC_{100}	~30-45	Shaw (1932)
Duck	oral	14-21 days	LC_{100}	~30	Shaw (1932)
	oral	> 15 days	LOEL	~50	Ward (1931)
	intraperitoneal	> 15 days	LOEL	~25	Ward (1931)
Cow	oral	14 days	LOEL	~25	Ward (1936)

[a] LC_{100} = concentration at which all animals were killed; LOEL = low-observed-effect level

Anthropogenic contamination, especially from cement plants in Germany, led to detailed studies of the effect of thallium-contaminated fodder on the development of farm animals. Continuous supplementation of maize-soybean fodder with 2, 4, 15 or 40 mg thallium(I) nitrate/kg in a 42-day broiler test and a 280(322)-day laying hen test caused obvious effects only at the highest concentration (Ueberschär et al., 1986). In comparison to control animals provided with uncontaminated feed, the body weight of the treated broilers and hens was reduced by about 14 and 32%, respectively; in the latter, laying rate (18%), feed efficiency (10%) and eggshell thickness (2%) were also reduced. In broilers fed with the two higher concentrations of thallium, gizzard erosion occurred.

In a detailed feeding study with fattening pigs with respect to performance, health and meat residues, low concentrations of thallium (daily intake of 0.05 and 0.1 mg thallium/kg body weight) were without any effects on weight gain, carcass quality, health, or haematological and biochemical parameters (Konermann et al., 1982). Daily administration of 0.3 or 1.0 mg/kg body weight in drinking-water was toxic to sheep, and the animals had to be killed after 4 and 6 weeks, respectively (Hapke et al., 1980). Administration of daily doses of 0.03 and 0.1 mg thallium/kg body weight to sheep (for 11 weeks in drinking-water) and 0.025 mg/kg body weight to steers (for 6 months in fodder) caused no deaths (Hapke et al., 1980). However, in both species daily uptake of 0.1 mg/kg body weight affected the animals after several weeks or months; fodder should therefore contain less than 0.5 mg/kg dry weight. Protein-rich food reduced the toxic action of thallium (Hapke, 1984).

10. EVALUATION

10.1 Evaluation of human health risks

10.1.1 Exposure levels

Since thallium is a naturally occurring element, humans are exposed to low levels in drinking-water, food and ambient air. Drinking-water concentrations are often below the level of detection (0.3 μg/litre) and rarely contribute more than 1 μg/litre. The total intake of thallium from drinking-water has been estimated to be < 1 μg/day for the vast majority of humans. In uncontaminated areas, the dietary contribution of thallium has been estimated to be less than 5 μg/day, with most of this coming from vegetables. Increased dietary intakes have been reported for individuals living in areas with thallium-contaminated soils; vegetables in these areas have been found to contain thallium concentrations 1-2 orders of magnitude higher than those grown in uncontaminated areas. However, the actual dietary intakes for individuals consuming contaminated vegetables have not been determined. In areas where there are no point sources of thallium, ambient air concentrations are very low (< 1 ng/m^3), typically contributing less than 0.005 μg/day to the total intake. Concentrations of thallium in workplace air can be several orders of magnitude higher than those in ambient air, resulting in a significantly increased total thallium intake. At the level of the threshold limit values (TLVs) in some countries (0.1 mg/m^3), the thallium intake from inhalation alone would be of the order of 1000 μg/day (assuming inhalation of 10 m^3 during a workshift). This intake from inhalation alone (which may be even higher in some workplaces) is about 500-fold higher than the total intake from non-occupationally exposed humans living in non-contaminated areas.

There are only limited data about the actual thallium content of workplace air. The most recent (1980s) concentrations of thallium observed were < 22 μg thallium/m^3 (in the production of a special thallium alloy and in a thallium smelter). Average urinary concentrations were determined to be in the range of 0.3-8 μg/litre for cement workers and 0.3-10.5 μg/litre for foundry workers.

10.1.2 Kinetics

Thallium is rapidly and well absorbed through the gastrointestinal and respiratory tracts and is also taken up through the skin. It is rapidly distributed to all organs and passes the placenta, as indicated by the rapid fetal uptake, and the blood-brain barrier. Because of its rapid accumulation in cells, concentrations of thallium in whole blood do not reflect the levels in tissues. In acute poisoning of experimental animals or humans, initially high concentrations of thallium appear in the kidney, low concentrations in fat tissue and brain, and intermediate concentrations in the other organs; later the thallium concentration of the brain also increases.

Elimination of thallium may occur through the gastrointestinal tract (mainly by mechanisms independent of biliary excretion), kidney, hair, skin, sweat and breast milk. Intestinal reabsorption (mainly from the colon) may occur with a consequent decrease in total body clearance. In rats, the main routes of thallium elimination are gastrointestinal (about 2/3) and renal (about 1/3), while in rabbits the contribution of the two routes is about equal. Thallium is also secreted in saliva.

As with many other substances, the excretion of thallium in humans differs from that in laboratory animals since the rate of excretion is generally much lower in humans (rate constant = 0.023-0.069 day^{-1}) than in animals (average rate constant = 0.18 day^{-1}). Another major difference between humans and animals is the relative contribution of the different routes of excretion. In humans, renal excretion seems to be much more important than in animals, although its relative contribution to the total body clearance has not been definitively established, due principally to the lack of sufficient human data. Moreover, exposure levels, duration of exposure, impairment of excretory organ function, potassium intake and concomitant treatment of acute poisoning may widely influence the results.

In a case report in which radioactive thallium (2.3 mg) was therapeutically administered, urinary excretion of thallium within 72 h after dosing was 11% of the administered concentration, whereas the gastrointestinal elimination was 0.5% during the same time period. According to this study, renal excretion of thallium is about 73%, whereas that through the gastrointestinal tract is about 3.7% of the daily excreted amount. Elimination through hair has been estimated to be 19.5% and that through skin and sweat 3.7%.

On the basis of the total daily excretion value, the daily intake of thallium has been estimated to be about 11 μg and 0.9 μg in chronically exposed and unexposed population, respectively; based on the total amount of thallium in the body, a daily intake of about 2.3 μg may be calculated in unexposed populations.

The biological half-life of thallium in laboratory animals generally ranges from 3 to 8 days. In humans it is about 10 days but values of up to 30 days have been reported.

No data on the biotransformation of thallium are available.

10.1.3 Toxic effects

Thallium salts are mainly tasteless, odourless, colourless and highly toxic. They were easily obtainable as rodenticides in the past and are still available in some developing countries. Acute thallium poisoning has resulted from accidental ingestion of thallium sulfate and its use for suicide, homicide and attempts at illegal abortion. Cases of homicide involving multiple low doses can induce chronic intoxication. Chronic thallium intoxication has been observed in occupationally exposed workers, and symptoms suggestive of thallium poisoning have been seen in population groups in contaminated areas.

Clinical manifestations of acute thallium poisoning may occur within hours or several days after exposure. Symptoms are often diffuse and initially may include anorexia, metallic taste, nausea, vomiting, retrosternal and abdominal pain, pain in the limbs, and paraesthesia. Gastrointestinal haemorrhage occasionally occurs; later on constipation is a common symptom. After the second day of thallium poisoning, effects on the central and peripheral nervous systems, skin, kidneys, eyes, cardiovascular and respiratory systems progressively develop. Extreme sensitivity and pain in the legs, later followed by the "burning feet" syndrome and paraesthesia, are common manifestations. Insomnia, depression, hallucination, lethargy, delirium, convulsions and coma may be followed by death, usually between 10 and 12 days. Where survival extends beyond a week or so, both motor and sensory neuropathy with cranial nerve involvement and retrobulbar neuritis may develop. Common circulatory disorders, such as hypertension, tachycardia and ischaemic cardiac changes, may also occur. Frequently loss of head hair and sometimes also body hair occurs after the second week of thallium poisoning. Dystrophy of the nails is manifested by the occurrence of lunular stripes (Mee's lines) 3 to 4 weeks after intoxication. Recovery requires months and occasionally some of the neurological and mental disturbances are permanent. Permanent blindness may follow retrobulbar neuritis and optic nerve atrophy.

Clinical features are generally milder in cases of chronic poisoning than in acute thallium intoxication. Occurrence of chronic thallium poisoning usually begins with neurological symptoms such as tiredness, fatigue, headache and insomnia. In some cases the first clinical findings include alopecia and constipation. The triad of gastroenteritis, polyneuropathy and alopecia is regarded as the classical syndrome of thallium poisoning, but in some cases gastroenteritis and alopecia have not been reported.

Postmortem examinations following thallium poisoning reveal damage in various organs. Haemorrhage in the mucosa of the intestine, lungs and heart, kidney damage, fatty infiltration of the liver and heart, and degeneration of neurons, including ganglion cells and axons, with disintegration of myelin sheaths have all been observed.

Limited data are available on the effects of thallium on human reproduction. Libido and male potency have been found to be adversely affected in poisoning cases. There is no adequate evidence for a genotoxic effect of thallium, and there have been no reports of any carcinogenic or immunological effects.

Following low-level environmental exposure to thallium, a dose-response relationship has been shown between thallium excretion in urine and the prevalence of tiredness, weakness, sleep disorders, nervousness, headache, muscle and joint pain and paraesthesia. Based on replies to a questionnaire, a similar dose-response relationship was seen when thallium in hair was taken as an indicator of exposure and uptake.

10.1.4 Dose-response relationship (animals)

No lifetime studies of thallium administration have been conducted on laboratory animals. In addition, no studies by the route of inhalation are available. Three studies of intermediate duration by the oral route are described in this report. A no-observed-effect level could not be determined from any study. The lowest doses were used in a 90-day gavage study (0, 0.01, 0.05 or 0.25 mg/kg body weight per day). Small but statistically significant changes in some clinical chemistry parameters were seen at the lowest dose level, as was alopecia. From animal studies, it therefore appears that an intake of 0.01 mg/kg per day may be associated with adverse effects. No doses lower than this have been tested.

On the basis of LC_{50} values in animals and known lethal doses in humans, it appears that humans may be more sensitive than laboratory rodents to the toxic effects of thallium. Because of the availability of human data and the apparently greater sensitivity of humans, a quantitative evaluation of animal data for use in a risk assessment has not been conducted here.

10.1.5 Dose-response relationship (humans)

Cases of acute thallium poisoning (with symptoms and signs listed in the section 10.1.3) have occurred as a result of ingestion of doses of thallium (in the form of soluble salts) as low as 1.5 mg/kg body weight. Higher doses give rise to more severe symptoms. Doses that have given rise to lethal poisoning are in the order of 10 mg/kg.

Concerning risks related to long-term exposure to lower doses of thallium, the Task Group considered that an evaluation, although uncertain, could best be performed on the basis of observed relationships between urinary excretion of thallium and the occurrence of symptoms. The urinary excretion value can be taken as an indicator of the daily total absorbed dose from inhalation and dietary intake.

A population-based study on unexposed healthy subjects living in northern Italy was performed with the aim of determining trace element concentrations, including thallium, in blood, serum or plasma, and urine, in which the collection, handling and analysis of the samples was carried out under rigorous standardized protocols. The 496 subjects in this study, drawn from both urban and rural areas were screened for normality by means of a questionnaire and clinical and biochemical examination (with the exclusion of those with a history of occupational exposure, heavy smokers, and those in a diseased state). The mean urinary thallium concentration was 0.42 ± 0.09 μg/litre (range 0.07-0.7 μg/litre). Other carefully controlled studies in population samples showed similar urinary concentrations, e.g., 0.4 ± 0.2 μg/litre and 0.3 ± 0.2 μg/litre in rural and urban population samples, respectively, and 0.3 ± 0.14 μg/litre in a sample of 149 subjects. This gives credence to a mean value of 0.3-0.4 μg/litre for urinary thallium concentration in an unexposed population. In all three studies, involving a total of 686 subjects, the range of urinary thallium concentrations was 0.06-1.2 μg/litre. As thallium has a short biological half-life, measured in days, and if a steady state can be assumed to exist in such population-based samples, the above urinary excretion value can be taken as an indicator of total dose in terms of absorption following inhalation and total daily dietary intake.

By contrast, in a population sample living in the vicinity of thallium emission into the atmosphere, the mean urinary thallium concentration was 5.2 μg/litre ± 8.3 μg/litre (range 0.1-76.5 μg/litre). Although a questionnaire on health effects was compiled on each subject, no objective tests were performed. From the replies to the questionnaire a clear dose-response relationship was found between thallium concentration in urine and the prevalence of tiredness, weakness, sleep disorder, headache, nervousness, paraesthesia, muscle and joint pain. A similar dose-response relationship was found when thallium in hair was taken as an indicator of exposure. In a limited study on cement plant workers with thallium exposure, where five workers showed urinary thallium levels above 5 μg/litre, but where the time interval between cessation of exposure and urine collection was not stated, paraesthesia was reported in five workers and distal muscle weakness in three. However, these symptoms could not be related to thallium exposure.

From the above limited studies it is suggested that an approximately 15-fold increase in urinary excretion of thallium above the mean non-exposed level of 0.3 to 0.4 μg/litre may be related to subjective symptoms which could possibly be considered as early adverse health effects.

It is known from clinical practice that there is an increased urinary concentration of thallium in acute poisoning cases. In 14 cases of thallium poisoning with recovery after therapy, the urinary thallium concentrations ranged from 500 to 20 400 μg/litre. In seven of these cases concentrations were below 2700 μg/litre. It should be recognized that these values are not entirely comparable to those in long-term exposure since they do not represent steady-state conditions.

In summary, the Task Group considered that exposures causing urinary thallium concentrations below 5 μg/litre are unlikely to cause adverse health effects. In the range of 5-500 μg/litre the magnitude of risk and severity of adverse effects are uncertain, while exposures giving values over 500 μg/litre have been associated with clinical poisoning.

10.2 Evaluation of the effects of thallium on the environment

Thallium is an element which occurs naturally in the earth's crust, primarily in the monovalent form. In marine water and some localized strongly oxidizing freshwater and soil, thallium may be present primarily in the oxidized trivalent form. The major anthropogenic sources of thallium released to the environment are smelting of metallic ores, mining, special cement production, and the combustion of fossil fuels, principally coal. Relatively little thallium is released into the environment because of the production and use of thallium compounds. Thallium levels reported in air are generally < 1 ng/m^3 although mean values up 15 ng/m^3 have been reported in industrial and urban air. Thallium may be released directly to the environment following its use as a rodenticide, although such use has been restricted or banned in many countries.

Thallium tends to persist in soil, although it may be leached to water under acidic conditions. Monovalent thallium is relatively stable in solution whereas trivalent thallium may be removed from the water column by precipitation as the oxide or hydroxide. Although thallium can bioconcentrate, it is unlikely to biomagnify in aquatic or terrestrial food webs. Thallium concentrations in water tend to be low, a maximum concentration of 2.4 mg/litre having been reported for industrial waste water. Thallium concentrations in surface water from industrial regions have been reported to range from 1 to 100 μg/litre, while surface water in uncontaminated areas normally contains lower levels. Concentrations of thallium in seawater range from < 0.01 to 0.02 μg/litre.

Most studies of effects on aquatic organisms have used soluble monovalent thallium compounds. Acute toxic effects have been reported in freshwater algae exposed to thallium at 100 μg/litre. Reduced growth of aquatic macrophytes was reported following a 28-day exposure to 8 μg/litre. The 48-h LC_{50} reported for *Daphnia* was 2200 μg/litre, while a 24-h LC_{50} of 110 μg/litre has also been reported. The 96-h LC_{50} values for freshwater fish range from 860 to 132 000 μg/litre. An LC_{50} of 40 μg/litre has been reported for freshwater fish exposed to thallium for approximately 40 days. The 96-h LC_{50} values for marine species are 2100 μg/litre for invertebrates and 20.9-24 mg/litre for fish. The available aquatic toxicity data suggest that thallium can harm aquatic organisms. However toxic effects are likely to be limited to sites adjacent to point sources such as some metal mining and smelting operations and cement plants.

Thallium concentrations in uncontaminated soil typically range from about 0.1 to 1.0 mg/kg dry weight, although higher levels can occur near natural sources such as thallium-enriched shales and some metallic ore deposits. Levels are generally somewhat elevated near anthropogenic sources such as cement plants using thallium-containing pyrite (up to 21 mg/kg dry weight) and base metal smelters (up to 2.1 mg/kg dry weight) that release large quantities of thallium to the atmosphere.

Very few data have been identified concerning the effects on terrestrial organisms of thallium in soil. The results of one study suggest that microbial community structure is disturbed at concentrations in the range of 1 to 10 mg/kg dry weight. However, the properties of the soil and the form of thallium used in this study were not identified.

Plants growing in uncontaminated soil normally contain 0.01 to 0.3 mg thallium/kg dry weight, while those growing near cement plants using thallium-enriched pyrite have been reported to contain much larger amounts (100 to 1000 mg/kg dry weight). Reduced growth has been reported in sensitive plant species at concentrations of about 1 mg thallium/kg of dry plant tissue following exposure to monovalent thallium. Toxic effects on terrestrial plants are therefore possible near some cement plants using thallium-enriched pyrite.

The use of thallium as a rodenticide has resulted in poisoning of non-target organisms including foxes, badgers, martens, partridges, pheasants and eagles. Poisoning of domestic animals, such as dogs, cats, ducks and pigeons, has also been widely reported. The number of wildlife poisoning incidents has declined as a result of the reduced use of thallium as a rodenticide. In countries with naturally high thallium levels, such as former-Yugoslavia and Israel, some farm animals have been poisoned following ingestion of vegetation with a high thallium content. Symptoms of thallium poisoning have been reported in cows that were calculated to have consumed 0.75 mg thallium/kg body weight per day for a 6-week period.

11. CONCLUSIONS AND RECOMMENDATIONS

The currently limited industrial uses of thallium are unlikely to pose a threat to the general environment. At industrial facilities such as metal mining and smelting operations and cement plants using pyrite, which can release significant amounts of thallium, the concentration of thallium in industrial raw materials as well as stack gases and waste water should be monitored and, if necessary, controlled. Waste materials containing water-soluble thallium compounds should be sealed and marked to avoid leaching and pollution by dust.

In the general population, environmental exposure to thallium does not pose a health threat. The total intake has been estimated to be less than 5 μg/day, with the vast majority coming from foodstuffs; drinking-water and air generally contribute very small amounts of thallium.

Due to its toxicity to both humans and non-target environmental species, the use of thallium as a rodenticide has been prohibited in many countries. Where thallium is still available for such use, however, the potential for accidental poisoning or for its use in homicide or suicide remains a significant concern. It is recommended that the use of thallium as a rodenticide be prohibited worldwide, particularly as less hazardous methods of rodent control are available.

Atmospheric emissions from industrial sources (e.g., cement plants using thallium-containing pyrite) have resulted in increased concentrations of thallium in biological samples (e.g., urine and hair) from the population living in the vicinity. A relationship was found between thallium concentrations in urine and hair and the prevalence of symptoms possibly indicating early health effects of thallium. The limited available data are not sufficient for determining an acceptable limit for emissions. Steps should be taken, however, to limit emissions to the greatest extent possible. Where thallium may be released into the environment, monitoring of both atmospheric emissions and resulting dust deposition rates should be performed. Where environmental monitoring reveals thallium levels significantly above background, it is recommended that biomonitoring of the population living in the vicinity of the point source be carried out. If biomonitoring reveals excessive exposure to thallium, emissions from the point source should be re-evaluated and an effort made to reduce them.

Since current occupational exposures to thallium may be of concern to health, it is recommended that measures be taken to reduce occupational exposure (as described in section 8.7 of this monograph). Furthermore, revision of threshold limit values for thallium warrants consideration.

All thallium analyses should be accompanied by a quality assurance programme. This requires certified reference materials of one matrix and of a similar concentration range to the sample to be analysed and participation in an inter-laboratory comparison programme. There is a need to make such reference materials available.

Since thallium is rapidly and well absorbed and its excretion is mainly renal, concentrations of thallium in urine may be considered a relatively reliable indicator of exposure. Exposure to thallium causing urinary concentrations below 5 µg/litre is unlikely to cause adverse human health effects. For thallium exposure giving rise to urinary concentrations in the range 5-500 µg/litre, the magnitude of risk and the severity of adverse effects on human health are uncertain, while exposure giving rise to 500 µg/litre or more has been associated with clinical poisoning. The estimated daily oral intake corresponding to a urinary thallium concentration of 5 µg/litre is approximately 10 µg thallium in the form of a soluble compound.

In view of the considerable uncertainties in the evaluation, the Task Group concluded that it was not possible to recommend a health-based exposure limit. Until better information on the dose-response relationship becomes available, it seems prudent to keep exposures at levels that lead to urinary concentrations of less than 5 µg/litre.

12. FURTHER RESEARCH

a) Follow-up epidemiological studies of populations exposed chronically to high levels of thallium (e.g., in the vicinity of cement plants and natural sources of high concentrations of thallium) should be performed to determine whether there is an increased risk of pathological effects, e.g., cancer, effects on reproduction (especially sperm cells) and congenital malformations. Such follow-up studies are also required in order to assess objectively the largely subjective symptom complex experienced by certain populations in thallium-contaminated areas.

b) Toxicokinetics of thallium, with particular respect to distribution and excretion, should be studied in people with low-level environmental exposure (in both uncontaminated and contaminated areas) and in those exposed occupationally. Adequate information from acutely poisoned patients should also be compiled to determine toxicokinetics and the relationship with the clinical findings. In addition, an attempt should be made to correlate clinical findings with thallium concentrations in biological samples. Patients should be followed for several years to ascertain the long-term consequences of acute poisonings.

c) Early indicators of an effect of thallium on glomerular or tubular kidney function, on liver function, on haem biosynthesis and on nerve conduction velocity should be sought in asymptomatic population samples and in occupationally exposed workers where there is evidence of excessive thallium exposure.

d) Early indicators of an effect of thallium absorption should be sought. The effects of thallium on the haem biosynthetic pathway suggest that estimation of porphyrins in blood and/or urine may be a useful approach to detecting specific early effect of thallium cellular toxicity. Similarly observations on the endoplasmic reticulum structure with respect to protein synthesis and polysome function suggest a need for further studies on liver function.

e) Chronic bioassays in laboratory animals are needed to assess the carcinogenic activity of thallium at low doses.

f) Lifetime studies in laboratory animals should be conducted using several low doses by the routes of both inhalation and ingestion to help determine a dose-response relationship and a no-observed-effect level.

g) Multi-generation studies of thallium in laboratory animals are needed to ascertain the reproductive and teratogenic effects of thallium.

h) Sample collection, analysis and data presentation should be carried out according to a protocol which ensures adequate validation of biological monitoring procedures.

i) The concentrations of thallium in environmental compartments adjacent to point sources should be determined using current analytical techniques. The bioavailability of thallium from various matrices should also be determined.

j) Total dietary studies should be conducted in areas with high thallium concentrations (either natural or from industrial sources) to determine the oral intake in populations living in these regions.

k) Thallium concentrations in soil and vegetables from regions with extensive acid rain should be determined.

REFERENCES

ACGHI (1980) Documentation of the threshold limit values, 4th ed. Cincinnati, Ohio, American Conference of Governmental Industrial Hygienists.

Achenbach C, Ziskoven R, Koehler F, Bahr U, & Schulten H-R (1979a) [Quantitative trace analysis of thallium in biological material.] Angew Chem, **91**: 944-945 (in German).

Achenbach C, Hauswirth O, Heindrichs C, Ziskoven R, Köhler F, Smend J, & Kowalewski S (1979b) [Toxicity and teratogenicity of thallium.] Dtsch Ärztebl, **76**: 3189-3192 (in German).

Achenbach C, Hauswirth O, Heindrichs C, Ziskoven R, Köhler F, Bahr U, Heindrichs A, & Schulten H-R (1980) Quantitative measurement of time-dependent thallium distribution in organs of mice by field desorption mass spectrometry. J Toxicol Environ Health, **6**: 519-528.

Achenbach C, Wiemer J, Ziskoven R, & Winter U (1982) Tl^{+}-Ions: effects on the automaticity of sinoatrial tissue and on dV/dt_{max} and i_{K2} of cardiac Purkinje fibres. Z Naturforsch, **37c**: 1006-1014.

Adamson RH, Canellos GP, & Sieber SM (1975) Studies on the antitumor activity of gallium nitrate (NSC-15200) and other group IIIa metal salts. Cancer Chemother Rep, **59**: 599-610.

Aderjan R, von Clarmann M, Daldrup T, Geldmacher-von Mallinckrodt M, van Heijst ANP, Machbert G, Schütz H, & Stöppler M (in press) [Thallium - clinical toxicological analysis after acute intoxications. Communications of the Senate Committee on clinical and toxicological analysis.] German Research Council (in German).

Agnes G, Hill HAO, Pratt JM, Ridsdale SC, Kennedy FS, & Williams RJP (1971) Methyl transfer from methyl vitamin B_{12}. Biochim Biophys Acta, **252**: 207-211.

Agrawal OP & Khatkar VB (1988) Amperometric determination of Ga(III), In(III), Tl(I) and Pb(II) with thiomalic acid. J Electrochem Soc India, **37**: 289-292.

Aitio A (1988) Biological monitoring. In: Clarkson T, Friberg L, Nordberg G, & Sager P ed. Biological monitoring of toxic metals. New York, London, Plenum Press, pp 75-83.

Alarcón-Segovia D, Amigo MC, & Reyes PA (1989) Connective tissue disease features after thallium poisoning. J Rheumatol, **16**: 171-174.

Al-Attar AF, Martin MH, & Nickless G (1988) Uptake and toxicity of cadmium, mercury and thallium to *Lolium perenne* seedlings. Chemosphere, **17**: 1219-1225.

Albertano P & Pinto G (1986) The action of heavy metals on the growth of three acidophilic algae. Boll Soc Nat Napoli, **95**: 319-328.

Ali SL & Blume H (1983) [Determination of the pesticide residues and toxic heavy metal traces in crude drugs.] Pharm Ztg, **128**: 2248-2252 (in German).

Ali SF, Jairaj K, Newport GD, Lipe GW, & Slikker W (1990) Thallium intoxication produces neurochemical alterations in rat brain. Neurotoxicology, **11**: 381-390.

Allus MA, Martin MH, & Nickless G (1987) Comparative toxicity of thallium to two plan species. Chemosphere, **16**: 929-932.

Allus MA, Brereton RG, & Nickless G (1988) Chemometric studies of the effect of toxic metals on plants: the use of response surface methodology to investigate the influence of Tl, Cd and Ag on the growth of cabbage seedlings. Environ Pollut, **52**: 169-181.

Andersen O (1984) Clinical evidence and therapeutic indications in neurotoxicology, exemplified by thallotoxicosis. Acta Neurol Scand, **70**(suppl 100): 185-192.

André T, Ullberg S, & Winqvist G (1960) The accumulation and retention of thallium in tissues of the mouse. Acta Pharmacol Toxicol, **16**: 229-234.

Anschütz M, Herken R, & Neubert D (1981) Studies on embryotoxic effects of thallium using the whole-embryo culture technique. In: Neubert D & Merker H-J ed. Culture techniques: applicability for studies on prenatal differentiation and toxicity. Proceedings of the 5th Symposium on Prenatal Development. Berlin, De Gruyter, pp 57-66.

Aoyama H (1989) Distribution and excretion of thallium after oral and intraperitoneal administration of thallous malonate and thallous sulfate in hamsters. Bull Environ Contam Toxicol, **42**: 456-463.

Aoyama H, Yoshida M, & Yamamura Y (1988) Induction of lipid peroxidation in tissues of thallous malonate-treated hamster. Toxicology, **53**: 11-18.

Apostoli P, Maranelli G, Minoia C, Massola A, Baldi C, & Marchiori L (1988) Urinary thallium: critical problems, reference values and preliminary results of an investigation in workers with suspected industrial exposure. Sci Total Environ, **71**: 513-518.

Arndt U, Finkbeiner M, Maier-Reiter W, & Wystrcil H-G (1987) [Ecotoxicology of thallium. Behaviour and location in phytocenosis.] Agrar-Umweltforsch Baden-Württemberg, **16**: 5-21 (in German).

Arnold W (1986) [Excretion and distribution of thallium in human poisoning.] In: Anke M, Baumann W, Bräunlich H, Brückner C, & Groppel B ed. [5th Symposium of Trace Elements, Jena, 14-17 July 1986.] Jena, University of Jena, pp 1241-1253 (in German).

Atkins HL, Budinger TF, Lebowitz E, Ansari AN, Greene MW, Fairchild RG, & Ellis KJ (1977) Thallium-201 for medical use. Part 3: Human distribution and physical imaging properties. J Nucl Med, **18**: 133-140.

ATSDR (1992) Toxicological profile for thallium. Atlanta, Georgia, Agency for Toxic Substances and Disease Registry, pp 1-90 (ATSDR/TP-91/26).

Aulenbach DB, Meyer MA, Beckwith E, Joshi S, Vasudevan C, & Clesceri NL (1987) Removal of heavy metals in POTW. Environ Prog, **6**: 91-98.

Auler H (1926) [Therapy of cancer.] Z Krebsforsch, **23**: 153-156 (in German).

Bahiga LM, Kotb NA, & El-Dessoukey EA (1978) Neurological syndromes produced by some toxic metals encountered industrially or environmentally. Z Ernähr.wiss, **17**: 84-88.

Bank WJ, Pleasure DE, Suzuki K, Nigro M, & Katz R (1972) Thallium poisoning. Arch Neurol, **26**: 456-464.

Barbier F (1974) Treatment of thallium poisoning. Lancet, **2**(7886): 965.

Barckow J & Jenss H (1976) [Thallium intoxication. Combined treatment with haemodialysis, forced diuresis and antidote.] Med Klin, **71**: 1377-1382 (in German).

Barclay RK, Peacock WC, & Karnofsky DA (1953) Distribution and excretion of radioactive thallium in the chick embryo, rat, and man. J Pharm Exp Ther, **107**: 178-187.

Barlow SM & Sullivan FM (1982) Thallium and its compounds. In : Reproductive hazards of industrial chemicals: An evaluation of animal and human data. New York, Academic Press, chapter 41, pp 530-537.

Barnes MP, Murray K, & Tilley PJ (1984) Neurological deficit more than thirty years after chronic thallium intoxication (letter). Lancet, **1**: 562.

Barrach H-J & Neubert D (1985) The moderating influence of potassium on the thallium effect in tissue differentiation *in vitro*. Teratology, **31**: 42A.

Barrera H & Gómez-Puyou A (1975) Characteristics of the movement of K^+ across the mitochondrial membrane and the inhibitory action of Tl^+. J Biol Chem, **250**: 5370-5374.

Barroso-Moguel R, Ríos-Castañeda C, Villeda-Hernández J, Méndez-Armenta M, & Galván-Arzate S (1990) Neurotoxicity of thallium: biochemical and morphological study of organic lesions. Arch Invest Med, **21**: 115-122.

Bazzaz FA, Carlson RW, & Rolfe GL (1974) The effect of heavy metals on plants: Part I. Inhibition of gas exchange in sunflower by Pb, Cd, Ni and Tl. Environ Pollut, **7**: 241-246.

Bean JR & Hudson RH (1976) Acute oral toxicity and tissue residues of thallium sulfate in golden eagles, *Aquila chrysaetos*. Bull Environ Contam Toxicol, **15**: 118-121.

Bell W, Hendry RA, & Annett HE (1925) The specific action of lead on the chorion epithelium of the rabbit, contrasted with the action of copper, thallium and thorium. J Obstet Gynaecol Br Emp, **32**: 1-16.

Bendl BJ (1969) Thallium poisoning. Report of a case successfully treated with dithizone. Arch Dermatol, **100**: 443-446.

Berndt H & Sopczak D (1987) [Error recognition and programme optimization in ET-AAS by time-resolved signals (direct determination of element traces in urine: Cd,Co, Cr, Ni, Tl, Pb).] Fresenius Z Anal Chem, **329**: 18-26 (in German).

Bertram HP, Kemper FH, Müller C, & Cuellar JA (1985) Trace metal patterns of human blood, saliva and hair: Environmental specimen banking. Proceedings of the 5th International Conference on Heavy Metals in the Environment, Toronto, vol 2, pp 102-104.

Bettinelli M, Baroni U, & Pastorelli N (1988) Determination of arsenic, cadmium, lead, antimony, selenium and thallium in coal fly ash using the stabilised temperature platform furnace and Zeeman-effect background correction. J Anal At Spectrom, **3**: 1005-1011.

BGA (1979) [Thallium.] Berlin, The Federal Environment Agency (Report No. 63/28/7180) (in German).

Bienvenu P, Nofre C, & Cier A (1963) Toxicité générale comparée des ions métalliques. Relation avec la classification périodique. C R Acad Sci (Paris), **256**: 1043-1044.

Birge WJ (1978) Aquatic toxicology of trace elements of coal and fly ash. In: Proceeding of the Symposium on Energy and Environmental Stress in Aquatic Systems, Augusta Georgia, 2-4 November 1977, pp 219-240.

Bock KD, Merguet P, Schley G, Schümann HJ, Rausch-Stroomann J-G, Hocevar V Schröder E, & Murata T (1968) [Pathogenesis of hypertension and tachycardia in the acute thallium poisoning and acute intermittent porphyria.] Dtsch Med Wochenschr, **93** 2119-2124 (in German).

Bonnet EFP & Pedace EA (1979) [Forensic anatomical histopathology of thallium poisoning.] Prensa Med Argent, **66**: 29-35 (in Spanish).

Bornhausen M & Hagen U (1984) Operant behavior performance changes in rats after prenatal and postnatal exposure to heavy metals. Int Res Commun Syst Med Sci, **12** 805-806.

Bösche J & Magureanu I (1983) [Heavy metal burden in Nordbaden by thallium and cadmium - Results of investigations of kidneys.] In: Barz J, Bösche J, Frohberg H, Joachim H, Käppner R, & Mattern R ed. [Advances in forensic medicine. Festschrift for Georg Schmidt.] Berlin, Springer Verlag, pp 270-275 (in German).

Bowen HJM (1966) Trace elements in biochemistry. New York, London, Academic Press.

Bowen HJM (1979) Environmental chemistry of the elements. New York, London, Academic Press.

Boysen P (1992) [Heavy metals and other substances in fertilizers - Environmental research plan of the Federal Minister for Environment, Nature Conservation and Reactor Safety, research report 107 01 016/1.] Berlin, Erich Schmidt Verlag (Report No. UBA-FB 92-104) (in German).

Bradley-Moore PR, Lebowitz E, Greene MW, Atkins HL, & Ansari AN (1975) Thallium-201 for medical use. II: biologic behavior. J Nucl Med, **16**: 156-160.

Briese C & Nessler F (1985a) [Historical data on the discovery of thallium and its first applications.] Wiss Z Humboldt-Univ (Berlin) Math-Naturwiss Reihe, **34**: 750-751 (in German).

Briese C & Nessler F (1985b) [Therapy of thallium poisoning.] Wiss Z Humboldt-Univ (Berlin) Math-Naturwiss Reihe, **34**: 786-788 (in German).

Briese C, Nessler F, & Liebscher R (1985) [Production of thallium and its significance in the different political economies.] Wiss Z Humboldt-Univ (Berlin) Math-Naturwiss Reihe, **34**: 756-760 (in German).

Bringmann G & Kühn R (1982) [Results of toxic action of water pollutants on *Daphnia magna* tested by an improved standardized procedure.] Z Wasser Abwasser Forsch, **15**: 1-6 (in German).

Britten JS & Blank M (1968) Thallium activation of the (Na^+-K^+)-activated ATPase of rabbit kidney. Biochim Biophys Acta, **159**: 160-166.

Brockhaus A, Dolgner R, Ewers U, Wiegand H, Freier I, Jerman E, & Krämer U (1980) Excessive thallium absorption among a population living near a thallium emitting cement plant. In: Holmstedt B, Lauwerys R, Mercier M, & Robberfroid M ed. Mechanisms of toxicity and hazard evaluation. Proceedings of the 2nd International Congress on Toxicology, Brussels, 1980. Dev Toxicol Environ Sci, **8**: 565-568.

Brockhaus A, Dolgner R, Ewers U, Freier I, Jermann E, & Wiegand H (1981a) Repeated surveillance of exposure to thallium in a population living around a cement plant emitting thallium containing dust. In: Proceedings of the International Conference on Heavy Metals in the Environment, Amsterdam, September 1981. Luxembourg, Commission of the European Communities, pp 482-485.

Brockhaus A, Dolgner R, Ewers U, Krämer U, Soddemann H, & Wiegand H (1981b) Intake and health effects of thallium among a population living in the vicinity of a cement plant emitting thallium containing dust. Int Arch Occup Environ Health, **48**: 375-389.

Brodersen HP, Theilmeier A, Korsten S, Arendt U, Larbig D, & Reis HE (1992) Acute poisoning by a potentially lethal dose of thallium: quantitative comparison of different methods of elimination. Intensiv Notfallbehandl, **17**: 204-207.

Brown BE (1982) The form and function of metal-containing "granules" in invertebrate tissues. Biol Rev, **57**: 621-667.

Brown BT & Rattigan BM (1979) Toxicity of soluble copper and other metal ions to *Elodea canadensis*. Environ Pollut, **20**: 303-314.

Brown DR, Callahan BG, Cleaves MA, & Schatz RA (1985) Thallium induced changes in behavioral patterns: correlation with altered lipid peroxidation and lysosomal enzyme activity in brain regions of male rats. Toxicol Ind Health, **1**: 81-98.

Brumsack H-J (1977) Potential metal pollution in grass and soil samples around brickworks. Environ Geol, **2**: 33-41.

Brumsack H, Heinrichs H, & Lange H (1984) West German coal power plants as sources of potentially toxic emissions. Environ Technol Lett, **5**: 7-22.

Budavari S ed. (1989) The Merck index: An encyclopedia of chemicals, drugs, and biologicals, 11th ed. Rahway, New Jersey, Merck & Co.

Bulavintseva OA & Bulavintsev AG (1982) [Histoenzymic characteristics in rats chronically poisoned with thallium sulfate.] In: Sedov KR ed. [Geochemical environment: The problem of the health of the population of the BAM railway zone], pp 68-69 (in Russian).

Burger A & Starke K (1969) [Influence of thallium on the ATPases of the amine storing granules from the adrenal medulla and splenic nerves.] Experientia (Basel), **25**: 578-579 (in German).

Buschke A (1900) [Experiments on alopecia.] Berl Klin Wochenschr, **37**: 1235-1236 (in German).

Buschke A (1903) [Further experiments on alopecia and the localization of dermatoses.] Berl Klin Wochenschr, **40**: 888-890 (in German).

Buschke A (1911) [Clinical and experimental observations of alopecia (hypotrichosis) congenita.] Arch Dermatol Syph, **108**: 27-40 (+ plates IV and V) (in German).

Buschke A (1929) [Thallium.] Dtsch Med Wochenschr, **55**: 1546-1547 (in German).

Buschke A & Berman L (1927) [The chemical and biological relation of thallium and lead. Klin Wochenschr, **6**: 2428-2429 (in German).

Buschke A & Curth W (1928) [Therapeutical experiments with thallium, apart from its us against fungal diseases of the hair of the head.] Derm Z, **53**: 113-117 (in German).

Buschke A & Jacobsohn F (1922) [Experiments on the action of thallium.] Dtsch Me Wochenschr, **48**: 859-860 (in German).

Buschke A & Langer E (1927) [The forensic and the industrial-hygienic importance o thallium.] Münch Med Wochenschr, **74**: 1494-1497 (in German).

Buschke A & Peiser B (1922a) [The effect of thallium on the endocrine system.] Kli Wochenschr, **1**: 995 (in German).

Buschke A & Peiser B (1922b) [Experimental observations of the influence of thallium on the endocrine system.] Med Klin, **18**: 731-732 (in German).

Buschke A & Peiser B (1922c) [The endocrine effect and the practical importance of thallium.] Dermatol Wochenschr, **74**: 443-445 (in German).

Buschke A & Peiser B (1926) [Thallium alopecia and sensory hairs.] Klin Wochenschr, **5**: 977-979 (in German).

Buschke A, Zondek B, & Berman L (1927a) [The inhibitory effect of thallium on the estrous cycle of mice.] Klin Wochenschr, **6**: 683-685 (in German).

Buschke A, Christeller E, & Loewenstein L (1927b) [Alterations of the skull by experimental chronic thallium poisoning.] Klin Wochenschr, **6**: 1088-1089 (in German).

Buschke A, Löwenstein L, & Joel W (1928) [Further histological results of experimental chronic thallium poisoning.] Klin Wochenschr, **7**: 1515 (in German).

Byczkowski JZ & Sorenson JRJ (1984) Effects of metal compounds on mitochondrial function: a review. Sci Total Environ, **37**: 133-162.

Byrne C, Balasubramanian R, Overton EB, & Albert TF (1985) Concentrations of trace metals in the bowhead whale. Mar Pollut Bull, **16**: 497-498.

Canterford GS (1980) Formation and regeneration of abnormal cells of the marine diatom *Ditylum brightwelli* (West) Grunow. J Mar Biol Assoc (UK), **60**: 243-254.

Canterford GS & Canterford DR (1980) Toxicity of heavy metals to the marine diatom *Ditylum brightwelli* (West) Grunow: correlation between toxicity and metal speciation. J Mar Biol Assoc (UK), **60**: 227-242.

Careaga-Olivares J & Morales-Aguilera A (1990) The distribution of thallous ion between blood cells and plasma at equilibrium. Arch Invest Med (Mex), **1990**: 21-23.

Careaga-Olivares J & Morales-Aguilera A (1993) Differential accumulation of thallous ion by diverse rabbit and rat muscles. Bull Environ Contam Toxicol, **51**: 764-771.

Carlson RW, Bazzaz FA, & Rolfe GL (1975) The effect of heavy metals on plants. II. Net photosynthesis and transpiration of whole corn and sunflower plants treated with Pb, Cd, Ni, and Tl. Environ Res, **10**: 113-120.

Cataldo DA & Wildung RE (1978) Soil and plant factors influencing the accumulation of heavy metals by plants. Environ Health Perspect, **27**: 149-159.

Cavanagh JB (1979) Metallic toxicity and the nervous system. In: Smith WT & Cavanagh JB ed. Recent advances in neuropathology. Edinburgh, Churchill Livingstone, pp 247-275.

Cavanagh JB (1988) Neurotoxic effects of metals and their interactions. Recent Adv Nerv Syst Toxicol, **100**: 177-202.

Cavanagh JB (1991) Annotation. What have we learnt from Graham Frederick Young? Reflections on the mechanism of thallium neurotoxicity. Neuropathol Appl Neurobiol, **17**: 1-9.

Cavanagh JB & Gregson M (1978) Some effects of thallium salts on the proliferation of hair follicle cells. J Pathol, **125**: 179-191.

Cavanagh JB, Fuller NH, Johnson HRM, & Rudge P (1974) The effects of thallium salts, with particular reference to the nervous system changes. Q J Med New Ser, **43**: 293-319.

Chandler HA & Scott M (1986) A review of thallium toxicology. J R Navy Med Serv, **72**: 75-79.

Chandler HA, Archbold GPR, Gibeon JM, O'Callaghan P, Marks JN, & Petghybridge RJ (1990) Excretion of a toxic dose of thallium. Clin Chem, **36**: 1506-1509.

Chattopadhyay A & Jervis RE (1974) Multielement determination in market-garden soils by instrumental photon activation analysis. Anal Chem, **46**: 1630-1639.

Christensen HE, Luginbyhl TT, Hill RH Jr, McGrath DJ, Georgevich M, Mitchell FL, & May JR ed. (1973) The toxic substances list, 1973 edition. Washington, DC, US Department of Health, Education, and Welfare, Public Health Service, US Government Printing Office.

Chusid JG & Kopeloff LM (1962) Epileptogenic effects of pure metals implanted in motor cortex of monkeys. J Appl Physiol, **17**: 697-700.

Clausen B & Karlog O (1974) Loading with thallium among wild animals of the marten genus and badgers in Denmark. Nord Vet.med, **26**: 339-350.

Claussen U, Roll R, Dolger R, Matthiaschk G, Majewski F, Stoll B, & Röhrborn G (1981) [Mutagenicity and teratogenicity of thallium - with special consideration of the situation in Lengerich.] Rhein Ärztebl, **16**: 469-475 (in German).

Cooper GP, Suszkiw JB, & Manalis RS (1984) Presynaptic effects of heavy metals. In: Narahashi T ed. Cellular and molecular neurotoxicology. New York, Raven Press, pp 1-21.

Cornelis R (1988) Analytical procedures and clinical reference materials in monitoring human exposures to trace metals with special reference to Cr, Pb and Tl. Sci Total Environ, **71**: 269-283.

Cortes Toro E, Parr RM, & Clements SA (1990) Biological and environmental referenc materials for trace elements, nuclides and organic microcontaminants. Vienn International Atomic Energy Agency (IAEA/RL/128 Rev.1).

Cottier H (1980) [Pathogenesis.] Berlin, Springer Verlag (in German).

Cotton FA & Wilkinson G (1988) Advanced inorganic chemistry, 5th ed. New York Chichester, Brisbane, Toronto, John Wiley and Sons.

Coyle V (1980) Diagnosis and treatment of thallium toxicosis in a dog. J Small Anim Prac **21**: 391-397.

Craig PJ (1980) Metal cycles and biological methylation. In: Hutzinger O ed. The handboo of environmental chemistry. Volume 1: The natural environment and the biogeochemica cycles - Part A. Berlin, Springer Verlag, pp 169-227.

Cromartie E, Reichel WL, Locke LN, Belisle AA, Kaiser TE, Lamont TG, Mulhern BM Prouty RM, & Swineford DM (1975) Residues of organochlorine pesticides an polychlorinated biphenyls and autopsy data for bald eagles, 1971-72. Pestic Monit J, 9 11-14.

Crössmann G (1984) [Thallium - a new environmental contamination?] Angew Bot, **58** 3-10 (in German).

Crössmann G (1985) [Toxic elements in feed.] Dtsch Tierärztl Wochenschr, **92**: 232-233 (in German).

Csicsaky M & Wiegand H (1981) [Neurotoxical effects of thallium on the peripheral an central nervous systems of the rat.] Umwelthygiene, **13**: 180 (in German).

Csicsaky M, Wiegand H, Uhlig S, Lohmann H, & Papadopoulos R (1988) Miniatur endplate potentials as a tool in neurotoxicology. Toxicology, **49**: 121-129.

Curry AS, Grech JL, Spiteri L, & Vassallo L (1969) Death from thallium poisoning: A case report. Eur J Toxicol, **2**: 260-269.

Danilewicz M, Danilewicz M, & Kurnatowski A (1979) [Dynamics of histopathological changes in the kidney of experimental animals after lethal doses of thallium.] Med Pracy, **30**: 257-265 (in Polish).

Danilewicz M, Kurnatowski A, & Wagrowska-Danilewicz M (1980) [Dynamics of pathomorphological changes in some internal organs of rats poisoned with lethal doses of thallous sulfate and treated with dithiocarb, with special attention to renal changes.] Patol Pol, **31**: 557-565 (in Polish).

Date AR, Cheung YY, & Stuart ME (1988) Application of inductively coupled plasma mass spectrometry to the analysis of iron ores. J Anal At Spectrom, **3**: 653-658.

Davids P, Güthner G, Lange M, Lohrer W, & Vahrenholt F (1980) [Situation and perspectives of the techniques for purification of air.] Staub-Reinhalt Luft, **40**: 392-402 (in German).

Davis RD, Beckett PHT, & Wollan E (1978) Critical levels of twenty potentially toxic elements in young spring barley. Plant Soil, **49**: 395-408.

Davis LE, Standefer JC, Kornfeld M, Abercrombie DM, & Butler C (1981) Acute thallium poisoning: toxicological and morphological studies of the nervous system. Ann Neurol, **10**: 8-44.

Dawson GW, Jennings AL, Drozdowski D, & Rider E (1975/77) The acute toxicity of 47 industrial chemicals to fresh and saltwater fishes. J Hazard Mater, **1**: 303-318.

De Groot G & van Heijst ANP (1988) Toxicokinetic aspects of thallium poisoning. Methods of treatment by toxin elimination. Sci Total Environ, **71**: 411-418.

Den Dooren de Jong LE & Roman WB (1971) Tolerance of *Azotobacter* for metallic and non-metallic ions. Antonie van Leeuwenhoek J Microbiol Serol, **37**: 119-124.

De Ruck A, Vandecasteele C, & Dams R (1987) Determination of thallium in natural waters by electrothermal atomic absorption spectrometry. Mikrochim Acta, **II**: 187-193.

De Ruck A, Vandecasteele C, & Dams R (1989) Determination of thallium in solid environmental samples by electrothermal atomic absorption spectrometry. Anal Lett, **22**: 69-479.

DFG (German Research Council) (1990) [Maximal concentrations at the place of work and biological tolerance values of material for work 1990.] Weinheim, Verlag Chemie (in German).

Di Gaudio MR & Hirshfield HI (1976) Respiratory toxicity of thallium in *Paramecium bursaria*. J Cell Biol, **70**: 60A.

Dilling WJ & Healey CW (1926) Influence of lead and the metallic ions of copper, zinc, thorium, beryllium and thallium on the germination of frogs' spawn and on the growth of tadpoles. Ann Appl Biol, **13**: 177-188.

Dolgner R & Wiegand H (1982) [Biological monitoring of thallium - medical aspects.] Staub-Reinhalt Luft, **42**: 151-152 (in German).

Dolgner R, Brockhaus A, Ewers U, Wiegand H, Majewski F, & Soddemann H (1983) Repeated surveillance of exposure to thallium in a population living in the vicinity of a cement plant emitting dust containing thallium. Int Arch Occup Environ Health, **52**: 79-94.

Douglas KT, Bunni MA, & Baindur SR (1990) Thallium in biochemistry. Int J Biochem, **22**: 429-438.

Downs WL, Scott JK, Steadman LT, & Maynard EA (1960) Acute and sub-acute toxicity studies of thallium compounds. Am Ind Hyg Assoc J, **21**: 399-406.

Drucker H, Garland TR, & Wildung RE (1979) Metabolic response of microbiota to chromium and other metals. In: Kharasch N ed. Trace metals in health and disease. New roles of metals in biochemistry, the environment, and clinical/nutritional studies. New York, Raven Press, pp 1-25.

Ducket S, Hiller D, & Ballas SK (1983) Quantitation and localization of ^{204}thallium in the central and peripheral nervous system of adult and young rats. Neurotoxicology, **4**: 227-234.

Dumitru D & Kalantri A (1990) Electrophysiologic investigation of thallium poisoning. Muscle Nerve, **13**: 433-437.

Durbin PW, Scott KG, & Hamilton JG (1957) The distribution of radioisotopes of som heavy metals in the rat. Berkeley, University of California Press, pp **3**: 1-34 (Universi of California Publications in Pharmacology, Volume 3).

Dvorák P (1970) [Hexacyanoferrates (II) as thallium antidotes. Preparation and properties Arzneimittelforschung, **20**: 1886-1888 (in German).

Dvorák P (1973) [Removal by rhyolitic pumice of internally deposited thallium from th rat.] Res Exp Med, **162**: 63-66 (in German).

Dvornikov AG, Ovsyannikov LB, & Sidenko OG (1973) [Biogeochemical dispersion halo of chalcophyllite elements in gold ore manifestations of the Nagolnyi Ridge (The Donbas) Dopov Akad Nauk Ukr RSR, **B35**: 490-494 (in Russian).

Dvornikov AG, Ovsyannikova LB, & Sidenko OG (1976) [Some characteristics of th coefficients of biological absorption and the biogeochemical coefficients at hydrotherm deposits of the Donets Basin in relation to predicting hidden ore mineralization Geokhimiya, **4**: 626-633 (in Russian).

Ebarvia B, Macalalad E, & Rogue N (1988) Determination of silver, cadmium, seleniu tellurium and thallium in geochemical exploration samples by atomic absorptio spectrometry. J Anal At Spectrom, **3**: 199-203.

Edel Rade J, Marafante E, Sabbioni E, Gregotti C, Di Nucci A, & Manzo L (198 Placental transfer and retention of ^{201}Tl-thallium in the rat. Toxicol Lett, **11**: 275-280.

EC (1987) [Legal regulations on hazardous substances.] Luxembourg, Office for Offici Publications of the European Community, vol 2 (in German).

Ehrhardt K (1927) [The effect of maternal intoxication with thallium on the progeny Klin Wochenschr, **6**: 1374-1375 (in German).

Emara M & Soliman MA (1950) The distribution of ingested thallium in the tissues o animals. J Egypt Med Assoc, **33**: 1-15.

Emsley J (1978) The trouble with thallium. New Sci, **79**: 392-394.

Erbslöh J (1960) [Thallium poisoning in the second half of pregnancy.] Arch Toxicol, **18** 156-159 (in German).

Ewan RC (1978) Toxicology and adverse effects of mineral imbalance with emphasis o selenium and other minerals. In: Oehme FW ed. Toxicity of heavy metals in th environment. New York, Marcel Dekker Inc., vol 1, pp 445-489.

Ewers U (1988) Environmental exposure to thallium. Sci Total Environ, **71**: 285-292.

Ewers U & Brockhaus A (1982) [Biological monitoring of thallium - analysis and results. Staub-Reinhalt Luft, **42**: 149-151 (in German).

Favari L & Mourelle M (1985) Thallium replaces potassium in activation of the (Na K^+)-ATPase of rat liver plasma membranes. J Appl Toxicol, **5**: 32-34.

Feudell P (1982) [Thallium polyneuropathology.] Psychiatr Neurol Med Psychol, **34**: 23-2 (in German).

scher AB (1981) Comparative studies of heavy metal toxicity at the cellular level. In: oceedings of the 3rd International Conference on Heavy Metals in the Environment, ronto, pp 486-490.

tzek A & Henning A (1976) [Distribution of thallium in the body of a pregnant cat.] sch Tierärztl Wochenschr, **83**: 66-68 (in German).

anjak J & Hodda AE (1988) Simultaneous determination of thallium, silver and gold in ine by flame and graphite-furnace atomic absorption spectrometry after extraction of e iodide complexes as ion-pairs with tri-*n*-octylamine. Anal Chim Acta, **207**: 283-289.

rd JK, Eyring EJ, & Anderson CE (1968) Thallium chondrodystrophy in chick embryos. n histological and biochemical investigation. J Bone Jt Surg, **50A**: 687-700.

rmigli L, Grégotti C, Di Nucci A, Edel J, Sabbioni E, & Manzo L (1985) Reproductive xicity of thallium. Experimental and clinical data. Proceedings of the 5th International onference on Heavy Metals in the Environment, Toronto, vol 2, pp 15-18.

rmigli L, Scelsi R, Poggi P, Grégotti C, Di Nucci A, Sabbioni E, Gottardi L, & Manzo (1986) Thallium-induced testicular toxicity in the rat. Environ Res, **40**: 531-539.

rth W & Henning CH (1979) [Thallium poisoning and the therapy.] Dtsch Ärztebl, **76**: 03-2807 (in German).

wler BA (1988) Mechanism of indium, thallium, and arsine gas toxicity. In: Clarkson W, Friberg L, Nordberg GF, & Sager PR ed. Biological monitoring of toxic metals. New ork, London, Plenum Press, pp 469-478.

rerking H, Osterkamp A, & Lindfeld KA (1990) [Adverse effects in cattle due to ccination against foot-and mouth disease or poisoning by thallium.] Dtsch Tierärztl ochenschr, **97**: 206-207 (in German).

alba J (1982) [Analysis of trace concentrations of phytotoxic elements in solid emissions om the thermal power plant ENO in soil and plants.] Pol'nohospodárstvo, **28**: 293-306 n Czechoslovakian).

alvan-Arzate S & Rios C (1994) Thallium distribution in organs and brain regions of eveloping rats. Toxicology, **90**: 63-69.

arcia Bugarin M, Casas JB, Sordo J, & Filella M (1989) Thallium(I) interactions in iological fluids: a potentiometric investigation of thallium(I) complex equilibria with some ulfur-containing amino acids. J Inorg Biochem, **35**: 95-105.

astel B (1978) Clinical conferences at the Johns Hopkins hospital: thallium poisoning. ohns Hopkins Med J, **142**: 27-31.

efel A, Liron M, & Hirsch W (1970) Chronic thallium poisoning. Isr J Med Sci, **6**: 80-382.

ehring PJ & Hammond PB (1967) The interrelation between thallium and potassium in nimals. J Pharmacol Exp Ther, **155**: 187-201.

eilmann W & Neeb KH (1959) [Application of the evaporation analysis for minor mounts of material. II. The proof and determination of minor contents of thallium.] resenius Z Anal Chem, **165**: 251-268 (in German).

Geilmann W, Beyermann K, Neeb K-H, & Neeb R (1960) [Thallium, a regularly occurri trace element in animals and plants.] Biochem Z, **333**: 62-70 (in German).

Geldmacher-von Malinckrodt M, Stamm D, & Enders P (1984) Interlaboratory survey thallium in urine. Z Rechtsmed, **93**: 219-225.

Ghezzi R & Bozza Marrubini M (1979) Prussian blue in the treatment of thalliu intoxication. Vet Hum Toxicol, **21**: 64-66.

Gibson JE & Becker BA (1970) Placental transfer, embryotoxicity, and teratogenicity thallium sulfate in normal and potassium-deficient rats. Toxicol Appl Pharmacol, **1** 120-132.

Gibson JE, Sigdestad CP, & Becker BA (1967) Placental transport and distribution thallium-204 sulfate in newborn rats and mice. Toxicol Appl Pharmacol, **10**: 378-411.

Glomme J (1983) Thallium and compounds. In: Encyclopaedia of occupational health a safety, 3rd ed. Geneva, International Labour Office, vol 2, pp 2170-2171.

Gluskoter HJ, Ruch RR, Miller WG, Cahill RA, Dreher GB, & Kuhn JK (1977) Tra elements in coal: occurrence and distribution - Final report. Urbana, Illinois, Illinois Sta Geological Survey (EPA 600/7-77-064).

Goenechea S & Sellier K (1967) [The natural concentration of thallium in humans.] Dtsc Z Gesamte Gerichtl Med, **60**: 135-141 (in German).

Goldblatt D (1989) Pollutants and industrial hazards. In: Rowland LP ed. Merritt's textboc of neurology, 8th ed. Philadelphia, Pennsylvania, Lea and Febinger, pp 919-928.

Gorbauch H, Rump HH, Alter G, & Schmitt-Henco CH (1984) [Determination of thalliu in raw materials and environmental samples.] Fresenius Z Anal Chem, **317**: 236-240 (German).

Gough LP, Shacklette HT, & Case AA (1979) Element concentrations toxic to plant animals, and man. US Geol Surv Bull, **1466**: 1-80.

Graben N, Doss M, & Klöppel HA (1978) [Disturbances of porphyrin metabolism b thallium poisoning.] Med Klin (Munich), **73**: 1114-1116 (in German).

Graben N, Klöppel H-A, Heidemann H, & Weiler G (1980) [Therapy of thalliu poisoning with special consideration of pregnancy and lactation.] Med Welt, **31**: 1391-139 (in German).

Grégotti C, di Nucci A, Formigli L, Sabbioni E, & Manzo L (1985) Altered testicula enzyme patterns in rats after long-term exposure to thallium sulfate. J Toxicol Clin Exp **5**: 265-271.

Grégotti C, Ponce R, & Faustman EM (1993) Mechanism of toxicity of Tl, B and Pb o testicular cells *in vitro*. Abstracts of the 32nd Annual Meeting of the Society of Toxicology Toxicologist, **13**: 298.

Gregus Z & Klaassen CD (1986) Disposition of metals in rats: a comparative study of feca urinary, and biliary excretion and tissue distribution of eighteen metals. Toxicol App Pharmacol, **85**: 24-38.

reving R & Gagel O (1928) [Polyneuritis after acute thallium poisoning.] Klin 'ochenschr, **7**: 1323-1325 (in German).

riepink B, Sager M, & Tölg G (1988) Determination of traces of thallium in various atrices. Pure Appl Chem, **60**: 1425-1436.

ubernator K, Moroni W, Munder S, Franke B, Ruske B, & Teufel D (1979) [The nbiguity of thallium contaminations in the neighbourhood of cement plants - Heidelberg eminar on Environmental Protection, Freiburg, Germany, October 1979.] Heidelberg, 'eidelberg Institute for Research on Energy and Environment, Institute for Applied cology (in German).

uderian R (1977) Air pollution. Berlin, Springer Verlag.

uidoboni RJ & Leipziger FD (1988) Glow discharge mass spectrometry - the newest tool or high purity materials analysis. J Cryst Growth, **89**: 16-20.

ünther M (1971) [Distribution and toxicity of thallium (I) in the rat as influenced by olloidal ferrihexacyanoferrate (II).] Arch Toxicol, **28**: 39-45 (in German).

ünther K & Umland F (1989) Bonding states of thallium and cadmium in thallium-reated and native rape. J Inorg Biochem, **36**: 63-74.

utknecht J (1983) Cadmium and thallous ion permeabilities through lipid bilayer nembranes. Biochim Biophys Acta, **735**: 185-188.

lagedorn-Götz H & Stoeppler M (1975) [The forensic proof of thallium in human hairs y flameless atomic absorption spectrometry.] Arch Toxicol, **34**: 17-26 (in German).

lajdu S & Nessler F (1985) [Chemistry, physics and occurrence of thallium and its ompounds.] Wiss Z Humboldt-Univ (Berlin) Math-Naturwiss Reihe, **34**: 752-755 (in German).

lall BK (1972) Thallium-induced achondroplasia in the embryonic chick. Dev Biol, **28**: 47-60.

lall BK (1977) Thallium-induced achondroplasia in chicken embryos and the concept of critical periods during development. Teratology, **15**: 1-15.

Hall BK (1985) Critical periods during development as assessed by thallium-induced nhibition of growth of embryonic chick tibiae *in vitro*. Teratology, **31**: 353-361.

Hamaguchi H (1960) [Inorganic constituents in biological materials. XIV. Thallium, selenium, and arsenic contents in Minamata districts.] Nippon Kagaku Zasshi, **81**: 920-927 (in Japanese).

Han HB, Kaiser G, & Tölg G (1982) Decomposition of biological materials, rocks sand soils with simultaneous volatilization of trace elements in pure oxygen under dynamic conditions. Anal Chim Acta, **134**: 3-11.

Hapke H-J (1984) [The effects of metals on domestic animals.] In: Merian E ed. [Metals in the environment.] Weinheim, Verlag Chemie, pp 181-194 (in German).

Hapke H-J, Barke E, & Spikermann A (1980) [Accumulation of thallium in the edib tissues of sheep and cattle according to the amount of thallium in the feed.] Dtsch Tierärz Wochenschr, **87**: 376-378 (in German).

Hart MM & Adamson RH (1971) Antitumor activity and toxicity of salts of inorgan group IIIa metals: aluminium, gallium, indium, and thallium. Proc Natl Acad Sci (USA **68**: 1623-1626.

Hart MM, Smith CF, Yancey ST, & Adamson RH (1971) Toxicity and antitumor activi of gallium nitrate and periodically related metal salts. J Natl Cancer Inst, **47**: 1121-112

Hasan M & Ali F (1981) Effects of thallium, nickel and cobalt administration on the lipi peroxidation in different regions of the rat brain. Toxicol Appl Pharmacol, **57**: 8-13.

Hasan M & Haider SS (1989) Acetyl-homocystein thiolactone protects against som neurotoxic effects of thallium. Neurotoxicology, **10**: 257-261.

Hasan M, Bajpai VK, & Shipstone AC (1977a) Electron microscopic study of thallium induced alterations in the oligodendrocytes of the rat area postrema. Exp Pathol, 1 338-345.

Hasan M, Chandra SV, Bajpai VK, & Ali SF (1977b) Electron microscopic effects c thallium poisoning on the rat hypothalamus and hippocampus: biochemical changes in th cerebrum. Brain Res Bull, **2**: 255-262.

Hasan M, Chandra SV, Dua PR, Raghubir R, & Ali SF (1977c) Biochemical an electrophysiologic effects of thallium poisoning on the rat corpus striatum. Toxicol Ap Pharmacol, **41**: 353-359.

Hasan M, Ashraf I, & Ali SF (1977d) Effect of thallium intoxication on the amino aci concentration of different regions of the rat brain. Curr Sci, **46**: 554-556.

Hasan M, Ashraf I, & Bajpai VK (1978a) Electron microscopic study of the effects o thallium poisoning on the rat cerebellum. Forensic Sci, **11**: 139-146.

Hasan M, Ali SF, & Tariq M (1978b) Levels of dopamine, norepinephrine an 5-hydroxytryptamine in different regions of the rat brain in thallium toxicosis. Act Pharmacol Toxicol, **43**: 169-173.

Hausman R & Wilson WJ jr (1964) Thallotoxicosis: a social menace. J Forensic Sci, 9 72-88.

Hayes WJ & Laws ER (1991) Handbook of pesticide toxicology - Volume 2: Classes o pesticides. New York, London, Academic Press.

Heath A, Ahlmén J, Branegård B, Lindstedt S, Wickström I, & Andersen O (1983 Thallium poisoning - Toxin elimination and therapy in three cases. J Toxicol Clin Toxicol **20**: 451-463.

Heinrichs H (1982) [Metals in sewage sludge from municipal water purification plants. Naturwissenschaften, **69**: 88-89 (in German).

Heinrichs H & Mayer R (1977) Distribution and cycling of major and trace elements ir two central European forest ecosystems. J Environ Qual, **6**(4): 402-407.

eit M (1985) Concentrations of potentially toxic trace elements in the muscle tissue of sh from acidic and circumneutral Adirondack Lakes. Proceedings of the 5th International onference on Heavy Metals in the Environment, Toronto, vol 1, pp 655-657.

enderson P, Hale TW, Shum S, & Habersang RW (1985) N-acetylcysteine therapy of ute heavy metal poisoning in mice. Vet Hum Toxicol, **27**: 522-525.

enke G (1991) Thallium determination in biological materials by radiochemical neuron tivation analysis. Fresenius J Anal Chem, **339**: 245-248.

ennekes R (1983) [Impairment of retinal function in acute thallium poisoning.] Klin lon.bl Augenheilkd, **182**: 334-336 (in German).

lenning CH & Forth W (1977) Absorption and secretion of thallous ions on isolated itestinal segments *in vitro* of rats. Naunyn-Schmiedeberg's Arch Pharmacol, **297**(suppl II): 2.

lenning CH & Forth W (1982) The excretion of thallium(I) ions into the gastrointestinal ract in situ of rats. Arch Toxicol, **49**: 149-158.

lenning CH, Becker G, & Forth W (1982) Movement of thallium(I) ions across the pithelium of intestinal segments *in vitro* of rats. Naunyn-Schmiedeberg's Arch Pharmacol, **21**: 157-163.

lenshaw JM, Heithmar EM, & Hinners TA (1989) Inductively coupled plasma mass pectrometric determination of trace elements in surface waters subject to acidic leposition. Anal Chem, **61**: 335-342.

lerman MM & Bensch KG (1967) Light and electron microscopic studies of acute and hronic thallium intoxication in rats. Toxicol Appl Pharmacol, **10**: 199-222.

Heydlauf H (1969) Ferric-cyanoferrate(II): an effective antidote in thallium poisoning. Eur Pharmacol, **6**: 340-344.

Heyl T & Barlow RJ (1989) Thallium poisoning: a dermatological perspective. Br J Dermatol, **121**: 787-791.

Hill WH & Murphy MA (1959) An appraisal of an industrial air contaminant by electrochromatography. Am Ind Hyg Assoc J, **20**: 387-391.

Hill HAO, Pratt JM, Ridsdale S, Williams FR, & Williams RJP (1970) Kinetics of substitution of co-ordinated carbanions in cobalt(III) corrinoids. Chem Commun, **124**: 341.

Hiltenkamp E & Jackwerth E (1988) [Investigations on the determination of Bi, Cd, Hg, Pb and Tl in high-purity gallium by graphite furnace AAS with atomization of metallic samples.] Fresenius Z Anal Chem, **332**: 134-139 (in German).

Hoffmann G, Schweiger P, & School W (1982) [Uptake of thallium by agriculturally and horticulturally useful plants.] Landwirtsch Forsch, **35**: 45-54 (in German).

Hollunger G (1960) The effect of trivalent thallium on oxidative phosphorylation. Acta Pharmacol Toxicol, **16**: 347-356.

Holm J, Brehmer R-D, Müller S, & Wester D (1987) [Bioindication of harmful substances taking roe deer and mallard as examples.] Fleischwirtschaft, **67**: 1145-1149 (in German).

Hologgitas J, Ullucci P, Driscoll J, Grauerholz J, & Martin H (1980) Thallium eliminatic kinetics in acute thallotoxicosis. J Anal Toxicol, **4**: 68-73.

Howe HE (1971) Chapter 27. Thallium. In: Hampel CA ed. Rare metals handboo Huntinston, New York, Huntinston Publishing Company, pp 529-535.

Hsie AW, Schenley RL, Tan E-L, Perdue SW, Williams MW, Hayden TL, & Turner J (1984) The toxicity of sixteen metallic compounds in Chinese hamster ovary cells: comparison with mice and *Drosophila*. Altern Methods Toxicol, **2**: 115-125.

Huber F, Schmidt U, & Kirchmann H (1978) Aqueous chemistry of organolead an organothallium compounds in the presence of microorganisms. In: Organometals an organometalloids: occurrence and fate in the environment). Washington, DC, America Chemical Society, pp 65-81 (ACS Symposium Series No. 82).

Hultin T & Näslund PH (1974) Effects of thallium(I) on the structure and functions o mammalian ribosomes. Chem-Biol Interact, **8**: 315-328.

Inturrisi CE (1969) Thallium activation of K^+-activated phosphatases from beef brai Biochim Biophys Acta, **173**: 567-569.

IPCS (International Programme on Chemical Safety) (1992) Poisons informatio monograph: Thallium. Geneva, World Health Organization (IPCS/INTOX/PIM.525).

IPS (Industrial Association for the Protection of Plants) ed. (1982) [Active compounds i agents for plant protection and pest control.] Offenbach, Bintz Verlag (in German).

Itawi RK & Turel ZR (1987) Determination of thallium in biological samples by therma neutron activation analysis and low-level beta-counting. J Radianal Nucl Chem, **115** 141-147.

Ivashchenko AT & Balmukhanov BS (1974) [Effect of monovalent ions on the adenosin triphosphatase activity of ascites tumor cells.] Izv Akad Nauk Kaz SSR Ser Biol, **12**: 73-78 (in Russian).

Ivashchenko AT, Balmukhanov BS, & Toktamysova ZS (1973) [Effects of monovalent ions on the glycolysis and respiration of ascite tumor cells.] Tsitologiya, **15**: 1487-1491 (in Russian).

Iyengar GV, Kollmer WE, & Bowen HJM (1978) The elemental composition of human tissues and body fluids. Weinheim, Verlag Chemie.

Jeantet AY, Ballan-Dufrançais C, & Martoja R (1977) Insect resistance to mineral pollution. Importance of spherocrystal in ionic regulation. Rev Ecol Biol Sol, **14**: 563-582.

Johnson CA (1976) The determination of some toxic metals in human liver as a guide to normal levels in New Zealand. Part I. Determination of Bi, Cd, Cr, Co, Cu, Pb, Mn, Ni, Ag, Tl and Zn. Anal Chim Acta, **81**: 69-74.

Junghans V & Nessler F (1985) [The clinical signs of chronic thallium intoxication shown in occupational thallium poisonings.] Wiss Z Humboldt-Univ (Berlin) Math-Naturwiss Reihe, **34**: 778-780 (in German).

Kaeser HE & Lambert EH (1962) Nerve function in experimental polyneuritis. Electroencephalogr Clin Neurophysiol Prog Electromyogr, **22**(suppl): 29-35.

Lamerbeek HH, Rauws AG, ten Ham M, & van Heijst ANP (1971a) Dangerous redistribution of thallium by treatment with sodium diethyldithiocarbamate. Acta Med Scand, **189**: 149-154.

Lamerbeek HH, Rauws AG, Ten Ham M, & van Heijst ANP (1971b) Prussian blue in therapy of thallotoxicosis: An experimental and clinical investigation. Acta Med Scand, **189**: 321-324.

Kaplan DI, Adriano DC, & Sajwan KS (1990) Thallium toxicity in bean. J Environ Qual, **19**: 359-365.

Karkos J (1971) [Neuropathological findings in thallium encephalopathy.] Neurol Neurochir Pol, **21**: 911-915 (in Polish).

Karnofsky DA & Lacon CR (1964) Effects of drugs on the skeletal development of the chick embryo. Clin Orthop Relat Res, **33**: 59-70.

Karnofsky DA, Ridgway LP, & Patterson PA (1950) Production of achondroplasia in the chick embryo with thallium. Proc Soc Exp Biol Med, **73**: 255-259.

Kazantzis G (1986) Thallium. In: Friberg L, Nordberg GF, & Vouk V ed. Handbook on the toxicology of metals, 2nd ed. Amsterdam, Oxford, New York, Elsevier Science Publishers, pp 549-567.

Kazantzis G (1994) Thallium and tin. In: Alessio L, Berlin A, Roi R, & van der Venne, M-Th ed. Biological indicators for the assessment of human exposure to industrial chemicals. Luxembourg, European Commission, Health and Safety Directorate, pp 41-92 (Report EUR 14815/EN).

Kemper FH (1979) [Thallium poisonings.] Münch Med Wochenschr, **121**: 1357-1358 (in German).

Kemper FH & Bertram HP (1984) [Thallium.] In: Merian E ed. [Metals in the environment.] Weinheim, Verlag Chemie, pp 571-583.

Kemper FH & Bertram HP (1991) Thallium. In: Merian E ed. Metals and their compounds in the environment. Weinheim, Verlag Chemie, pp 1227-1241.

Kennedy P & Cavanagh JB (1976) Spinal changes in the neuropathy of thallium poisoning. J Neurol Sci, **29**: 295-301.

Kijewski H (1984) Nature of the black zones in hair after ingestion of thallium. Arch Kriminol, **173**: 36-44.

Klahre P, Valenta P, & Nürnberg HW (1978) [A standardized pulse inverse voltametric analytical method for testing drinking water for toxic metals. Part 1: Simultaneous determination of copper, cadmium, lead and zinc, and lead and thallium.] Vom Wasser, **51**: 199-219 (in German)

Klöppel A & Weiler G (1978) [Contribution to heavy metal poisonings, especially by thallium.] Dtsch Med Wochenschr, **103**: 75-76 (in German).

Knapp G (1985) Sample preparation techniques - an important part in trace element analysis for environmental research and monitoring. Int J Environ Anal Chem, **22**: 71-83.

Kogan BI (1970) [Thallium.] Redk Elem, **5**: 27-48 (in Russian).

Kolb Meyers V (1983) Chemicals which cause birth defects - teratogens: a special concer of research chemists. Sci Total Environ, **32**: 1-12.

Konermann H, Crössann G, & Hoppenbrock KH (1982) [Investigations of the influenc of thallium-contaminated food rations on performance, health and meat residues i fattening pigs.] Tierärztl Umsch, **37**: 8-21 (in German).

Korkisch J & Steffan J (1979) Determination of thallium in natural waters. Int J Enviro Anal Chem, **6**: 111-118.

Kosta L (1982) Contamination as a limiting parameter in trace analysis. Talanta, **2** 985-992.

Krassowsky GN, Yurasova OI, Charyev OG, Skrypnikov AI, & Varshavskaya SP (1977 [Prognosis of the gonadotoxic action of heavy metals judging by the primary effect o material accumulation.] Gig i Sanit, **7**: 11-16 (in Russian).

Krassowsky GN, Wasjukowitsch LJ, Tscharyjew OG, & Kenessarijew UI (1980 [Experimental studies on making quick prognoses of a gonadotoxic effect of inorgani compounds.] Z Gesamte Hyg Grenzgeb, **26**: 336-338 (in German).

Kravzov J, Rios C, Altagracia M, Monroy-Noyola A, & López F (1993) Relationshi between physicochemical properties of Prussian blue and its efficacy as antidote agains thallium poisoning. J Appl Toxicol, **31**: 213-216.

Ku DD, Akera T, Oldaard MK, & Brody TM (1978) Effects of lithium and thallous ions on sodium. Activity in the guinea pig heart and their relationship to the positive inotropic action. Naunyn-Schmiedeberg's Arch Pharmacol, **304**: 167-173.

Kwan KHM & Smith S (1988) The effect of thallium on the growth of *Lemna minor* and plant tissue concentrations in relation to both exposure and toxicity. Environ Pollut, **52** 203-220.

Lameijer W & van Zwieten PA (1977a) Kinetic behaviour of thallium in the rat. Accelerated elimination of thallium owing to treatment with potent diuretic agents. Arch Toxicol, **37**: 265-273.

Lameijer W & van Zwieten PA (1977b) Kinetics of thallium elimination in the rat after acute and subacute intoxication; influence of diuretic treatment. Proc Eur Soc Toxicol, **18**: 163-164.

Lameijer W & van Zwieten PA (1979) The efficacy of a potassium-rich diet compared to diuretic treatment on the renal elimination of thallium in the rat. Arch Toxicol, **42**: 33-41.

Landauer W (1930) [Influence of thallium poisoning of the father on his descendants. Experiments on roosters.] Arch Gewerbepathol Gewerbehyg, **1**: 791-792 (in German).

Larson CP, Keller WN, & Manges JD (1939) Accidental canine thallotoxicosis and dangers of thallium used as a rodenticidal agent. J Am Vet Med Assoc, **95**: 486-489.

LeBlanc GA (1984) Interspecies relationships in acute toxicity of chemicals to aquatic organisms. Environ Toxicol Chem, **3**: 47-60.

eBlanc GA & Dean JW (1984) Antimony and thallium toxicity to embryos and larvae of ıthead minnows (*Pimephales promelas*). Bull Environ Contam Toxicol, **32**: 565-569.

ehmann FPA & Favari L (1984) Parameters for the adsorption of thallium ions by ctivated charcoal and Prussian blue. J Toxicol Clin Toxicol, **22**: 331-339.

ehmann FPA & Favari L (1985) Acute thallium intoxication: kinetic study of the relative fficacy of several antidotal treatments in rats. Arch Toxicol, **57**: 56-60.

ehn H & Bopp M (1987) [Heavy metals in soil and the determination of the availability or plants.] Angew Bot, **61**: 467-481 (in German).

eloux MS, Lich NP, & Claude JR (1987a) Flame and graphite furnace atomic absorption pectroscopy methods for thallium - a review. At Spectrom, **8**: 71-77.

eloux MS, Lich NP, & Claude JR (1987b) Experimental studies on thallium toxicity in ats. I. Localization and elimination of thallium after oral acute and sub-acute intoxication. Toxicol Clin Exp, **7**: 247-257.

e Quesne PM (1984) Toxic neuropathies. In: Asbury AK & Gilliatt RW ed. Butterworths nternational medical reviews: Neurology. Volume 4: Peripheral nerve disorders - a ractical approach. London, Butterworths, pp 184-204.

etourneau VA, Joshi BM, & Butler LC (1987) Comparison between Zeeman and ontinuum background correction for graphite furnace AAS on environmental samples. At Spectrom, **8**: 145-149.

evander OA & Argrett LC (1969) Effects of arsenic, mercury, thallium, and lead on elenium metabolism in rats. Toxicol Appl Pharmacol, **14**: 308-314.

iang Y, Shi J, Chen L, & Ding G (1980) [Dimercaptosuccinic acid per os promoted the xcretions of Pb, Cu, Sb, Sr, Tl and Pm.] Acta Pharm Sinica, **15**: 335-340 (in Chinese).

Lide DR (1990) CRC handbook of chemistry and physics: A ready reference book of chemical and physical data, 71st ed. Boca Raton, Florida, CRC Press.

Lie R, Thomas RG, & Scott JK (1960) The distribution and excretion of thallium-204 in the rat, with suggested MPC's and a bio-assay procedure. Health Phys, **2**: 334-340.

Lillie WI & Parker HL (1932) Retrobulbar neuritis due to thallium poisoning. J Am Med Assoc, **98**: 1347-1349.

Lindegren CC (1971) Mitochondria in intoxication and detoxication. Physiol Chem Phys, **3**: 499-500.

Lindegren CC & Lindegren G (1973a) Mitochondrial modification and respiratory deficiency in the yeast cell caused by cadmium poisoning. Mutat Res, **21**: 315-322.

Lindegren CC & Lindegren G (1973b) Oxidative detoxication of thallium in the yeast mitochondria. Antonie van Leeuwenhoek J Microbiol Serol, **39**: 351-353.

Ling GN (1977) Thallium and cesium in muscle cells compete for the adsorption sites normally occupied by K^+. Physiol Chem Phys, **9**: 217-225.

Linsdale JM (1931) Facts concerning the use of thallium in California to poison roden - Its destructiveness to game birds, song birds and other valuable wild life. Condor, 3 92-106.

Linton RW, Loh A, Natusch DFS, Evans CA Jr, & Williams P (1976) Surfa predominance of trace elements in airborne particles. Science, **191**: 852-854.

LIS (Land Institute for Protection against Emissions) (1980) [Environmental burden b thallium. Investigations in the neighbourhood of the Dyckerhoff cement plant in Lengeric and other thallium-emitting plants in the country NW.] Bonn, Bonn University Press (German).

Liu G (1986) [The protective effect of selenium on myocardial cells.] Chung-hua Ping I Hsueh Tsa Chih, **15**: 252-254 (in Chinese).

Lockwood TH (1974) The analysis of asbestos for trace metals. Am Ind Hyg Assoc J, 3 245-251.

Logan PG, Lepp NW, & Phipps DA (1983) Thallium uptake by higher plants. I Proceedings of the 4th International Conference on Heavy Metals in the Environmen Toronto, vol 1, pp 642-645.

Logan PG, Lepp NW, & Phipps DA (1984) Some aspects of thallium uptake by highe plants. In: Proceedings of the 18th Trace Substances in Environmental Health Conference Columbia, 4-7 June 1984, pp 570-575.

Lohmann H, Csissaky M, & Wiegand H (1989) The action of thallium on the excitabilit of CA1 pyramidal cells in hippocampal slices. Neurotoxicol Teratol, **11**: 545-549.

Luckey TD & Venugopal B (1977) Metal toxicity in mammals. Volume 1: Physiologic an chemical basis for metal toxicity. New York, Plenum Press.

Luckit J, Mir N, Hargreaves M, Costello C, & Gazzard B (1990) Thrombocytopeni associated with thallium poisoning. Hum Exp Toxicol, **9**: 47-48.

Ludolph A, Elger CE, Sennhenn R, & Bertram HP (1986) Chronic thallium exposure i cement plant workers: clinical and electrophysiological data. Trace Elem Med, **3**: 121-125

Ludwig E (1961) [Pathognomonic hair findings in thallium poisoning and thei significance.] Hautarzt, **12**: 456-459 (in German).

Lund A (1956a) Distribution of thallium in the organism and its elimination. Act Pharmacol Toxicol, **12**: 251-259.

Lund A (1956b) The effect of various substances on the excretion and the toxicity of thallium in the rat. Acta Pharmacol Toxicol, **12**: 260-268.

Lynch GR, Lond MB, & Scovell JMS (1930) The toxicology of thallium. Lancet, **219**(5599): 1340-1344.

Machtey I & Bandmann M (1961) The cardiovascular system in thallium poisoning. Proc Staff Meet Beilinson Hosp, **10**(suppl): 140-142.

McCool MM (1933) Effect of thallium sulphate on the growth of several plants and on nitrification in soils. Contrib Boyce Thompson Inst, **5**: 289-296.

McDonald JM (1941) Metallic poisons. Ind Med, **10**: 447-452.

McKillop A & Taylor EC (1973) Thallium in organic synthesis. Chem Br, **9**: 4-11.

McLaren JW, Beauchemin D, & Berman SS (1987) Application of isotope dilution inductively coupled plasma mass spectrometry to the analysis of marine sediments. Anal Chem, **59**: 610-613.

Mahoney W (1933) Further notes on retrobulbar neuritis due to thallium poisoning. Yale J Biol Med, **6**: 583-597.

Maier EA, Dietemann-Molard A, Rastegar F, Heimburger R, Ruch C, Maier A, Roegel E, & Leroy MJF (1986) Simultaneous determination of trace elements in lavage fluids from human bronchial alveoli by energy-dispersive x-ray fluorescence. 2. Determination of abnormal lavage contents and verification of the results. Clin Chem, **32**: 664-668.

Maier-Reiter W, Blind G, Klump A, Wystrcil HG, & Arndt U (1987) [Ecotoxicology of thallium. Effect on the pigment system of plants.] Agrar-Umweltforsch Baden-Württemberg, **16**: 22-36 (in German).

Makridis H & Amberger A (1989a) [Uptake of thallium from cement factory dust in pot trials with green rape, bush beans and rye grass.] Landwirtsch Forsch, **42**: 324-332 (in German).

Makridis H & Amberger A (1989b) [Effect of increasing levels of thallium on yield and uptake of thallium and further mineral nutrients by bush beans and green rape grown in nutrient solution]. Landwirtsch Forsch, **42**: 333-343 (in German).

Malcolm D (1979) Toxicology of metals other than lead. In: Gardner AW ed. Current approaches to occupational medicine. Bristol, John Wright and Sons Ltd., pp 18-43.

Manning DC & Slavin W (1988) The determination of thallium with the stabilized temperature platform furnace and Zeeman background correction. Spectrochim Acta, **43B**: 1157-1165.

Manzo L & Sabbioni E (1988) Thallium toxicity and the nervous system. In: Bondy SC & Prasad KN ed. Metal neurotoxicity. Boca Raton, Florida, CRC Press, pp 35-54.

Manzo L, Scelsi R, Moglia A, Poggi P, Alfonsi E, Pietra R, Mousty F, & Sabbioni E (1983) Long-term toxicity of thallium in the rat. In: Brown SS & Savory J ed. Chemical toxicology and clinical chemistry of metals. Proceedings of 2nd International Conference, Montreal, Canada. New York, London, Academic Press, pp 401-405.

Marcus RL (1985) Investigation of a working population exposed to thallium. J Soc Occup Med, **35**: 4-9.

Marwaha J, Freedman R, & Hoffer B (1980) Electrophysiological changes at a central noradrenergic synapse during thallium toxicosis. Toxicol Appl Pharmacol, **56**: 345-352.

Mason B (1966) Principles of geochemistry. New York, Chichester, Brisbane, Toronto, John Wiley and Sons.

Masri MS (1976) Binding of metallic ions to keratins. Proceedings of the 5th International Wool Textile Research Conference, vol 3, pp 1-12.

Matera MG, Marrazzo R, Berrino L, Fillippelli W, Russo S, Rossi F, Cilento M, & Marm E (1986) [Vascular effects of perinatal exposure of rats to thallium.] Rend Atti Accad S Med Chir, **140**: 177-202 (in Italian).

Mathis BJ & Kevern NR (1973) Distribution of mercury, cadmium, lead and thallium i a eutrophic lake. Washington, DC, US Department of Commerce, National Technic Information Service (NTIS PB-221993).

Mathis BJ & Kevern NR (1975) Distribution of mercury, cadmium, lead and thallium i a eutrophic lake. Hydrobiologia, **46**: 207-222.

Mathys W (1981) Thallium in river sediments of North-West Germany: its sources of inpu and its relation to other heavy metals. In: Proceedings of the International Conference o Heavy Metals in the Environment, Amsterdam, September 1981. Luxembourg, Commissio of the European communities WHO, pp 347-350.

Matthews AD & Riley JP (1969) The determination of thallium in silicate rocks, marin sediments and sea water. Anal Chim Acta, **48**: 25-34

Melnick RL, Monti LG, & Motzkin M (1976) Uncoupling of mitochondrial oxidativ phosphorylation by thallium. Biochem Biophys Res Commun, **69**: 68-73.

Merguet P, Schümann HJ, Murata T, Rausch-Stroomann JG, Schröder E, Parr D, & Bock KD (1969) [Investigations of the pathogenesis of hypertension and sinus tachycardia in human thallium poisonings.] Dtsch Arch Klin Med, **216**: 1-20 (in German).

Metter D & Vock R (1984) [Investigations of the structure of hairs in thallium poisoning.] Z Rechtsmed, **91**: 201-214 (in German).

Meyer H & Tal M (1957) Studies on thallium poisoning. I. The detoxifying action of some compounds containing sulphur and labile methyl groups. J Sci Food Agric, **8**: 516-519.

Micke H, Bertram HP, & Kemper FH (1983) [Thallium.] In: [Ullmann's encyclopedia of technical chemistry], 4th ed. Weinheim, Verlag Chemie, vol 23, pp 103-114 (in German).

Minoia C, Sabbioni E, Appostoli P, Pietra R, Pozzoli L, Gallorini M, Nicolaou G, Alessio L, & Capodaglio E (1990) Trace element reference values in tissues from inhabitants of the European Community I. A study of 46 elements in urine, blood and serum of Italian subjects. Sci Total Environ, **95**: 89-105.

MISA (Municipal/Industrial Strategy for Abatement) (1988/89) Preliminary report for the first six months of monitoring in the petroleum refining sector (December 1, 1988 to May 31, 1989). Ontario, Canada, Ministry of the Environment.

MISA (Municipal/Industrial Strategy for Abatement) (1991a) Status report on the effluent monitoring data for the iron and steel sector for the period from November 1, 1989 to October 31, 1990. Ontario, Canada, Ministry of the Environment.

MISA (Municipal/Industrial Strategy for Abatement) (1991b) Preliminary report on the first six months of process effluent monitoring in the MISA pulp and paper sector (January 1, 1990 to June 30, 1990). Ontario, Canada, Ministry of the Environment.

Moeschlin S (1965) [Poisoning: diagnosis and treatment.] Stuttgart, Georg Thieme (in German).

Moeschlin S (1986) [Poisoning: diagnosis and treatment.] Stuttgart, Georg Thieme (in German).

Möllhoff G, Schmidt G, & Bösche J (1979) [Thallium poisoning. Neuromedical aspects.] Arch Kriminol, **163**: 1-13 (in German).

Montoya Cabrera MA, Pérez Lucio C, Badillo Torre FF, Barquet Barquet RM, & López Bárcenas F (1979) [Thallium intoxication: treatment with D-penicillamine.] Rev. Med Inst Mex Seguro Soc, **18**: 211-214 (in Spanish).

Mottet NK (1985) Morphogenic injury by chemical agents. In: Mottet NK ed. Environmental pathology. New York, Oxford University Press, pp 82-125.

Mourelle M, Favari L, & Amezcua JL (1988) Protection against thallium hepatotoxicity by silymarin. J Appl Toxicol, **8**: 351-354.

Muller L (1961) [Thallium poisoning.] In: Baader EW ed. [Handbook of total occupational medicine.] Berlin, Urban & Schwarzenberg, vol II, part 1, pp 244-250.

Müller-Brand J, Staub JJ, & Friedrich R (1984) [Thallium scintigraphy of the thyroid as an alternative procedure for the demonstration of functionally silent thyroid tissue.] Schweiz Med Wochenschr, **114**: 1686-1689 (in German).

Mullins LJ & Moore RD (1960) The movement of thallium ions in muscle. J Gen Physiol, **43**: 759-773.

Munch JC (1934a) Antidotes. III. Thallium. J Am Pharm Assoc, **23**: 91-94.

Munch JC (1934b) Human thallotoxicosis. J Am Med Assoc, **102**: 1929-1934.

Munch JC, Ginsburg HM, & Nixon CE (1933) The 1932 thallotoxicosis outbreak in California. J Am Med Assoc, **100**: 1315-1319.

Munch B, Clausen B, & Karlog O (1974) Thallium poisoning in red foxes (*Vulpes vulpes*) and badgers (*Meles meles*) in Denmark. Nord Veterinärmed, **26**: 323-338.

Muramatsu Y & Parr RM (1985) Survey of currently available reference materials for use in connection with the determination of trace elements in biological and environmental materials. Vienna, International Atomic Energy Agency (IAEA/RL/128).

Nachman M & Hartley PL (1975) Role of illness in producing learned taste aversions in rats. A comparison of several rodenticides. J Comp Physiol Psychol, **89**: 1010-1018.

Naganuma A, Tanaka T, Maeda K, Matsuda R, Tabata-Hanyu J, & Imura N (1983) The interaction of selenium with various metals *in vitro* and *in vivo*. Toxicology, **29**: 77-86.

Natusch DFS (1982) Size distributions and concentrations of trace elements in particulate emissions from industrial sources. VDI-Ber, **429**: 253-260.

Natusch DFS & Wallace JR (1974) Urban aerosol toxicity: the influence of particle size. Science, **186**: 695-699.

Natusch DFS, Wallace JR, & Evans CA Jr (1974) Toxic trace elements: preferential concentration in respirable particles. Science, **183**: 202-204.

Negherbon WO (1959) Handbook of toxicology. Volume 3: Insecticides. Philadelphia Pennsylvania, W.B. Saunders Company, pp 750-751.

Nehring D (1962/63) [Investigations of the toxicological effects of thallium ions on fis and animal food of fish.] Z Fischerei, **11**: 557-561 (in German).

Nessler F (1985a) [Experimental studies made on animals to establish the toxicity o thallium and its compounds.] Wiss Z Humboldt-Univ (Berlin) Math-Naturwiss Reihe, **34** 761-766 (in German).

Nessler F (1985b) [Importance and history of occupational thallium intoxications as a typ of chronic thallium poisoning.] Wiss Z Humboldt-Univ (Berlin) Math-Naturwiss Reihe **34**: 773-777 (in German).

Nessler F & Briese C (1985) [Acute and subacute thallium intoxications of humans.] Wis Z Humboldt-Univ (Berlin) Math-Naturwiss Reihe, **34**: 767-772 (in German).

Neubert D & Bluth U (1985) Effect of thallium on limb development in organ culture Teratology, **32**: 29A.

NIOSH (1984) NIOSH manual of analytical methods: Method 7300.1-7300.5. Cincinnati Ohio, National Institute for Occupational Safety and Health.

NIOSH (1985/86) RTECS - Registry of toxic effects of chemical substances. Cincinnati Ohio, National Institute for Occupational Safety and Health, p 4637.

NIOSH (1992) NIOSH recommendations for occupational safety and health: Compendium of policy documents and statements. Cincinnati, Ohio, National Institute for Occupational Safety and Health (Publication No. 92-100).

Nisticò G, Faraone V, Naccari F, Ammendola D, Germanò D, & Marmo E (1984) Lack of effects of thallium on GABAergic mechanisms in discrete areas of the rat brain. Res Commun Chem Pathol Pharmacol, **46**: 35-42.

Noddack I & Noddack W (1939) [The frequency of heavy metals in sea organisms.] Ark Zool, **A32**(4): 1-35 (in German).

Nogami H & Terashima Y (1973) Thallium-induced achondroplasia in the rat. Teratology, **8**: 101-102.

Nolan C, Duke E, Lorenzon G, Sabbioni E, & Marafante E (1984) Heavy metal uptake and intracellular binding in isolated gill preparations of *Mytilus galloprovincialis* L. Sci Total Environ, **40**: 83-92.

Norris P, Man WK, & Hughes MN (1976) Toxicity and accumulation of thallium in bacteria and yeast. Arch Microbiol, **110**: 279-286.

Nriagu JO & Pacyna JM (1988) Quantitative assessment of worldwide contamination of air, water and soils by trace metals. Nature (Lond), **333**: 134-139.

Oehme FW (1972) Mechanisms of heavy metal toxicities. Clin Toxicol, **5**: 151-167.

Oehme FW (1978) Mechanisms of heavy metal inorganic toxicities. In: Oehme FW ed. Toxicity of heavy metals in the environment. New York, Marcel Dekker Inc., vol 1, pp 69-85.

Ohnesorge FK (1985) [Toxicological rating of arsenic, lead, cadmium, nickel, thallium and zinc.] Düsseldorf, VDI-Association of German Engineers Press (Progress Reports Series No. 15, Environmental Technology No. 38) (in German).

Olsen I & Jonsen J (1982) Whole-body autoradiography of ^{204}Tl in embryos, fetuses and placentas of mice. Toxicology, **23**: 353-358.

Ormerod MJ (1928) Pharmacological and toxicological aspects of thallium. Can Med Assoc J, **19**: 663-665.

Osorio-Rico L, Galván-Arzate S, & Ríos C (1994) Brain regional monoaminergic activity after acute thallium administration. Abstracts of the 33rd Annual Meeting of the Society of Toxicology. Toxicologist, **14**: 267.

Ostádalová I & Babicky A (1987) Protective effect of thallium against the formation of selenium cataract in rats. Biol Trace Elem Res, **14**: 101-103.

Overnell J (1975) The effect of some heavy metal ions on photosynthesis in a freshwater alga. Pestic Biochem Physiol, **5**: 19-26.

Pai V (1987) Acute thallium poisoning. Prussian blue therapy in 9 cases. West Indian Med J, **36**: 256-258.

Palermo D, Chiaravalle AE, D'Errico M, Milillo MA, & De Natale G (1983) [An occurrence of thallium poisoning in fish.] Clin Vet, **106**: 129-131 (in Italian).

Pallaghy CK (1972) Localization of thallium in stomata is independent of transpiration. Aust J Biol Sci, **25**: 415-417.

Papp JP, Gay PC, Dodson VN, & Pollard HM (1969) Potassium chloride treatment in thallotoxicosis. Ann Intern Med, **71**: 119-123.

Paschal DC & Bailey GG (1986) Determination of thallium in urine with Zeeman effect graphite furnace atomic absorption. J Anal Toxicol, **10**: 252-254.

Paschal DC, DiPietro ES, Phillips DL, & Gunter EW (1989) Age dependence of metals in hair in a selected U.S. population. Environ Res, **48**: 17-28.

Paulson G, Vergara G, Young J, & Bird M (1972) Thallium intoxication treated with dithizone and hemodialysis. Arch Intern Med, **129**: 100-103.

Peele DB, MacPhail RC, & Farmer JD (1986) Flavor aversions induced by thallium sulfate: importance of route of administration. Neurobehav Toxicol Teratol, **8**: 273-277.

Peele DB, Farmer JD, & MacPhail RC (1987) Conditioned flavor aversions: applications in assessing the efficacy of chelators in the treatment of heavy-metal intoxication. Toxicol Appl Pharmacol, **88**: 397-410.

Peisach M, Lackay LG, & Gihwala D (1989) Measurement of proton-induced prompt low energy photons by high resolution spectrometry. Analyst, **114**: 279-286.

Petersohn KL (1960) [Thallium poisonings in pregnancy.] Arch Toxikol, **18**: 160-164 (in German).

Peterson ER & Murray MR (1965) Patterns of peripheral demyelination *in vitro*. Ann NY Acad Sci, **122**: 39-50.

Piazolo P, Franz HE, Brech W, Walb D, & Wilk G (1971) [Treatment of thallium poisoning by hemodialysis.] Dtsch Med Wochenschr, **96**: 1217-1222 (in German).

Pielow E (1979) [Pollution by thallium in the environment of a cement plant.] Umwelt, **5**: 394-396 (in German).

Pieper B & Austenfeld F-A (1985) [Phytotoxicity of thallium (Tl) in culture solution. Part 2: Effects of Tl(III) on the growth and heavy metal contents of pea and field bean plants.] Z Pflanzenernähr Bodenkd, **148**: 83-91 (in German).

Potes-Gutierrez J & Del Real E (1966) Acute thallium intoxication. Ind Med Surg, **35**: 618-619.

Pötsch K & Austenfeld F-A (1985) [Phytotoxicity of thallium (Tl) in culture solution. Part 1: Effects of Tl(I) on the growth and heavy metal contents of pea and field bean plants.] Z Pflanzenernähr Bodenkd, **148**: 73-82 (in German).

Prick JJG (1979) Thallium poisoning. In: Vinken PJ & Bruyn GW ed. Handbook of clinical neurology: Intoxications of the nervous system. Amsterdam, North-Holland Publishers, vol 36, part 1, pp 239-278.

Prinz B, Krause GHM, & Stratmann H (1979) [Damages by thallium in the neighbourhood of the Dyckerhoff cement plant in Lengerich, Westfalen.] Staub-Reinhalt Luft, **39**: 457-462 (in German).

Puddu A, Pettine M, Bacciu F, La Noce T, & Pagnotta R (1985) Thallium and chromium toxicity in phytoplankton cultures. Proceedings of the 5th International Conference on Heavy Metals in the Environment, Toronto. vol 2, pp 304-306.

Puddu A, Pettine M, La Noce T, Pagnotta R, & Bacciu F (1988) Factors affecting thallium and chromium toxicity to marine algae. Sci Total Environ, **71**: 572.

Puerner NJ & Siegel SM (1972) The effects of mercury compounds on the growth and orientation of cucumber seedlings. Physiol Plant, **26**: 310-312.

Que Hee SS & Boyle JR (1988) Simultaneous multielemental analysis of some environmental and biological samples by inductively coupled plasma atomic emission spectrometry. Anal Chem, **60**: 1033-1042.

Rao DV, Govelitz GF, & Sastry KSR (1983) Radiotoxicity of thallium-201 in mouse testes: inadequacy of conventional dosimetry. J Nucl Med, **24**: 145-153.

Rauws AG & Canton JH (1976) Adsorption of thallium ions by Prussian Blue. Bull Environ Contam Toxicol, **15**: 335.

Reyners H, Gianfelici de Reyners E, Maisin JR, Winneke G, & Csicsaky M (1981) Effects of different heavy metals (Cd, Tl, Zn and Pb) in the central nervous system: a morphological assay. In: Proceedings of the International Conference on Heavy Metals in the Environment, Amsterdam, September 1981. Luxembourg, Commission of the European communities WHO, pp 495-497.

Richards OW (1932) The stimulation of yeast growth by thallium, a "bios" impurity of asparagine. J Biol Chem, **96**: 405-418.

Richelmi P, Bono F, Guardia L, Ferrini B, & Manzo L (1980) Salivary levels of thallium in acute human poisoning. Arch Toxicol, **43**: 321-325.

Ridgway LP & Karnofsky DA (1952) The effects of metals on the chick embryo: toxicity and production of abnormalities in development. Ann NY Acad Sci, **55**: 203-215.

Ridley WP, Dizikes LJ, & Wood JM (1977) Biomethylation of toxic elements in the environment. Science, **197**: 329-332.

Ríos C & Monroy-Noyola A (1992) D-penicillamine and Prussian blue as antidotes against thallium intoxication in rats. Toxicology, **74**: 69-76.

Ríos C, Galván-Arzate S, & Tapia R (1989) Brain regional thallium distribution in rats acutely intoxicated with Tl_2SO_4. Arch Toxicol, **63**: 34-37.

Ríos C, Kravsov J, Altagracia M, Lopez-Naranjo F, & Monroy A (1991) Efficacy of Prussian blue against thallium poisoning: effect of particle size. Proc West Pharmacol, **34**: 61-63.

Romero Romero E, Cherem JH, & Lifshitz A (1989) [Arterial hypertension in acute thallitoxicosis.] Rev Med Inst Mex Seguro Soc, **27**: 43-47 (in Spanish).

Rosenstock L & Cullen MR (1986) 9. Neurologic disease. In: Rosenstock L & Cullen MR ed. Clinical occupational medicine. Philadelphia, Pennsylvania, W.B. Saunders Company, pp 118-134.

Rossi F, Lampa E, Grella A, Vacca C, Giordano L, Perna D, Rosatti F, Spaziante G, Giasi M, Imperatore A, Giusti I, & Marmo E (1981) [Experimental study of the cardiovascular and respiratory effects of thallium in various animal species.] Rend Atti Accad Sci Med Chir, **134**: 49-84 (in Italian).

Rossi F, Marrazzo R, Berrino L, De Santis D, Lisa M, Susanna V, Montanaro C, Fici F, & Marmo E (1988) Prenatal and postnatal thallium exposure in rats: effect on development of vasomotor reactivity in pups. Teratog Carcinog Mutagen, **8**: 13-23.

Sabbioni E & Manzo L (1980) Metabolism and toxicity of thallium. In: Proceedings of the International Congress on Neurotoxicology, pp 249-270 (Advances in Neurotoxicology).

Sabbioni E, Marafante E, Rade J, Di Nucci A, Gregotti C, & Manzo L (1980a) Metabolic patterns of low and toxic doses of thallium in the rat. In: Holmstedt B, Lauwerys R, Mercier M, & Robberfroid M ed. Mechanisms of toxicity and hazard evaluation. Proceedings of the Second International Congress on Toxicoloy, Brussels, 1980. Amsterdam, Oxford, New York, Elsevier/North Holland, pp 559-564.

Sabbioni E, Goetz L, Marafante E, Gregotti C, & Manzo L (1980b) Metabolic fate of different inorganic and organic species of thallium in the rat. Sci Total Environ, **15**: 123-135.

Sabbioni E, Gregotti C, Edel J, Marafante E, Di Nucci A, & Manzo L (1982) Organ/tissue disposition of thallium in pregnant rats. Arch Toxicol, **5**(suppl): 225-230.

Sabbioni E, Di Nucci A, Edel J, Gregotti C, Marafante E, & Manzo L (1984a) Intestin absorption and excretion of thallium (^{201} Tl) in the rat. Arch Toxicol, **7**(suppl): 446-45

Sabbioni E, Goetz L, & Bignoli G (1984b) Health and environmental implication of trac metals released from coal fired power plants: an assessment study of the situation in th European Community. Sci Total Environ, **40**: 141-154.

Saddique A & Peterson CD (1983) Thallium poisoning: a review. Vet Hum Toxicol, **2** 16-22.

Sager M (1984) Separation and enrichment of traces of thallium. Mikrochim Acta, **198** 461-466.

Sager M (1986) [Trace element analysis of thallium.] In: Hulpke H, Hartkamp H, & Tö G ed. [Analytical chemistry for the practise.] Stuttgart, Georg Thieme, pp 1-103 (i German).

Sager M & Tölg G (1984) [Trace element analysis of thallium.] In: Fresenius W, Günzle H, Huber W, Lüderwald I, & Tölg G ed. [Analyst pocket book 4.] Berlin, Springer Verlag pp 443-466 (in German).

Sanderson L (1952) Thallium. Can Min J, **73**: 72-73.

Sanotskii IV (1961) [Toxicity of thallium compounds (carbonate, iodide, and bromide) Toksikol Novykh Prom Khim Veshchestv, **2**: 94-104 (in Russian).

Schafer EW (1972) The acute oral toxicity of 369 pesticidal, pharmaceutical and othe chemicals to wild birds. Toxicol Appl Pharmacol, **21**: 315-330.

Schäfer SG & Forth W (1980) Excretion of metals into the intestine; a comparative stud on rats. In: Proceedings of the Second International Congress on Toxicoloy, Brussels, 6-1 July 1980. Toxicol Lett, **95**(spec issue).

Schäfer SG & Forth W (1987) Thallium(I) secretion across the isolated mucosa of ra descending colon. Arch Toxicol, **52**: 413-420.

Schäfer SG, Nell G, & Henning CH (1981) Movement of thallium(I) ions *in vitro*. Arch Toxicol, **48**: 271-279.

Schaller K-H, Manke G, Raithel HJ, Bühlmeyer G, Schmidt M, & Valentin H (1980) Investigations of thallium-exposed workers in cement factories. Int Arch Occup Environ Health, **47**: 223-231.

Schardein JL & Keller KA (1989) Potential human developmental toxicants and the role of animal testing in their identification and characterization. Crit Rev Toxicol, **19** 251-339.

Scharrer K (1955) [Biochemistry of trace elements.] Berlin, Parey (in German).

Schaub GA & Schnitker A (1988) Influence of *Blastocrithidia triatomae* (Trypanosomatidae) on the reduviid bug *Triatoma infestans*: alterations in the Malpighian tubules. Parasitol Res, **75**: 88-97.

Schaumburg HH & Berger A (1992) Alopecia and sensory polyneuropathy from thallium in a chinese herbal medication. J Am Med Assoc, **268**: 3430-3431.

chmidbauer H & Klingler D (1979) [Chronical thallium poisoning.] Wien Med Vochenschr, **129**: 334-336 (in German).

choental R & Cavanagh JB (1977) Mechanisms involved in the 'dying-back' process - an ypothesis implicating coenzymes. Neuropathol Appl Neurobiol, **3**: 145-157.

choer J (1984) Thallium. In: Hutzinger O ed. The handbook of environmental chemistry. Volume 3: Anthropogenic compounds, Part C. Berlin, Springer Verlag, pp 143-214.

choer J & Nagel U (1980) [Thallium in plants and soils. Investigations in the southern Rhein-Neckar-Kreis.] Naturwissenschaften, **67**: 261-262 (in German).

choll W & Metzger F (1982) [Determinations of the thallium burden of useful plants on ontaminated soils in the region of Lengerich.] Landwirtsch Forsch, **35**: 216-223 (in German).

chwartz JG, Stuckey JH, Kunkel SP, Dowd DC, & Kagan-Mallet KS (1988) Poisoning rom thallium. Tex Med, **84**: 46-48.

chwartzman RM & Kirschbaum JO (1962) The cutaneous histopathology of thallium oisoning. J Invest Dermatol, **39**: 169-173.

Schweiger P & Hoffmann G (1983) [Thallium uptake of different plants in culture olution.] Kali-Briefe, **16**: 391-398 (in German).

Seeger R & Gross M (1981) [Thallium in higher fungi.] Z Lebensm.Unters Forsch, **173**: -15 (in German).

Segebade C & Schmitt BF (1987) Analysis of high-purity material - a comparison of photon activation analysis with other instrumental methods. J Radioanal Nucl Chem, **113**: 61-76.

Sessions HK & Goren S (1947) Report of investigation of health hazards in connection with the industrial handling of thallium. US Navy Med Bull, **47**: 545-550.

Shabalina LP & Spiridonova VS (1979) [Thallium as an industrial poison (Review of literature).] J Hyg Epidemiol Microbiol Immunol, **23**: 247-255 (in Russian).

Shabalina LP, Spiridonova VS, Pavlenko GI, & Fomenko VN (1980) [Skin resorptive, local, and mutagenic effects of thallium carbonate.] Gig Tr Prof Zabol, **4**: 54-55 (in Russian).

Shacklette HT, Erdmann JA, Harms TF, & Papp CSE (1978) Trace elements in plant foodstuffs. In: Oehme FW ed. Toxicity of heavy metals in the environment. New York, Marcel Dekker Inc., vol 1, pp 25-68.

Sharma RP & Obersteiner EJ (1981) Metals and neurotoxic effects: cytotoxicity of selected metallic compounds on chick ganglia cultures. J Comp Pathol, **91**: 235-244.

Sharma J, Sharma RL, Singh HB, & Satake M (1986) Hazards and analysis of thallium. A review. Toxicol Environ Chem, **11**: 93-116.

Shaw PA (1932) Studies on thallium poisoning in game birds. Calif Fish Game, **18**: 29-34.

Sherlock JC & Smart GA (1986) Thallium in foods and the diet. Food Addit Contam, **3**: 363-370.

Siegel SM (1977) The cytotoxic response of *Nicotiana*' protoplasts to metal ions: a surve of the chemical elements. Water Air Soil Pollut, **8**: 293-304.

Siegel BZ & Siegel SM (1976) Effect of potassium on thallium toxicity in cucumbe seedlings: further evidence for potassium-thallium ion antagonism. Bioinorg Chem, 341-345.

Sikka R, Sabherwal U, & Arora DR (1987) Susceptibility of *Klebsiella pneumoniae* t heavy metal ions *in vitro*. Indian J Med Res, **86**: 437-440.

Singh NP, Bodgen JD, & Joselow MM (1975) Distribution of thallium and lead i children's blood. Arch Environ Health, **30**: 557-558.

Skrovina B, Micek J, Hronská J, & Spissák L (1973) Experimental and huma achondroplasia. Teratology, **8**: 237.

Skulskii IA & Lapin AV (1983) Effect of thallium ions on sodium and potassium transpor in frog skin. Tsitologiya, **25**: 1284-1288.

Skulskii IA, Manninen V, & Järnefelt J (1973) Interaction of thallous ions with the catio transport mechanism in erythrocytes. Biochim Biophys Acta, **298**: 702-709.

Skulskii IA, Manninen V, & Järnefelt J (1975) Thallium inhibition of ouabain-sensitiv sodium transport and of the $(Na^{+}+K^{+})$-ATPase in human erythrocytes. Biochim Biophy Acta, **394**: 569-576.

Skulskii IA, Savina MV, Glasunov VV, & Saris NEL (1978) Electrophoretic transport o Tl^{+} in mitochondria. J Membr Biol, **44**: 187-194.

Skulskii IA, Ivanova TI, & Savina MV (1984) The effect of monovalent thallium salts o liver mitochondria of rats. Zh Evol Biokhim Fiziol, **20**: 353-355 (in Russian).

Smith IC & Carson BL (1977) Trace metals in the environment: Volume 1 - Thallium. Ann Arbor, Michigan, Ann Arbor Science Publishers, Inc.

Smith S & Kwan MKH (1989) Use of aquatic macrophytes as a bioassay method to asses relative toxicity, uptake kinetics and accumulated forms of trace metals. Hydrobiologia **188/189**: 345-351.

Sommer B (1987) [Evaluation of the effects of cadmium, lead, thallium and fluoride in plants for food and fodder - Environmental research plan of the Federal Minister for Environment, Nature Conservation and Reactor Safety, research report 106 07 049/02.] Berlin, Erich Schmidt Verlag (Report No. UBA-FB 87-026/8/87) (in German).

Spencer EL (1937) Frenching of tobacco and thallium toxicity. Am J Bot, **24**: 16-24.

Spiridonova VS, Shabalina LP, & Kremenetskaya LE (1978) [Fibrogenic activity of thallium halides.] Gig Tr Prof Zabol, **11**: 54 (in Russian).

Srivastava TN, Bajpai KK, & Singh K (1973) Anti-microbial activities of diaryl gallium, indium and thallium compounds. Indian J Agric Sci, **43**: 88-93.

Staszyc J, Królikowska-Prasal I, Matysiak W, Korobowicz E, Chróst L, & Kusmierczyk S (1986) Morphological evaluation of the lungs of rats affected by soil dust. Med Pr, **37**: 65-72 (in Polish).

avinoha WB, Emerson GA, & Nash JB (1959) The effects of some sulfur compounds on allotoxicosis in mice. Toxicol Appl Pharmacol, **1**: 638-646.

euer W (1980) [Results of thallium examinations of the human population in aden-Württemberg.] Öff Gesundh-Wesen, **42**: 477-479 (in German).

evens W, van Peteghem C, Heyndrickx A, & Barbier F (1974) Eleven cases of thallium toxication treated with Prussian blue. Int J Clin Pharmacol, **10**: 1-22.

okinger HE (1987) Thallium. In: Clayton GD & Clayton FE ed. Patty's industrial hygiene nd toxicology: Volume 2A: Toxicology, 3rd ed. New York, Chichester, Brisbane, Toronto, ohn Wiley and Sons, pp 1914-1931.

truempler AW (1975) Trace element composition in atmospheric particulates during 1973 nd the summer of 1974 at Chadron, Neb. Environ Sci Technol, **9**: 1164-1168.

underman FW (1967) Diethyldithiocarbamate therapy of thallotoxicosis. Am J Med Sci, **53**: 209-220.

wain RE & Bateman WG (1909/10) The toxicity of thallium salts. J Biol Chem, **7**: 37-151.

'ackmann W & Lehmann HJ (1971) [Refractory period and impulse sequence recorded rom the tibial nerves of guinea pigs with acute thallium polyneuropathy.] Z Neurol, **199**: 05-115 (in German).

'alas A & Wellhöner HH (1983) Dose-dependency of Tl^{+} kinetics as studied in rabbits. Arch Toxicol, **53**: 9-16.

'an E-L, Williams MW, Schenley RL, Perdue SW, Hayden TL, Turner JE, & Hsie AW 1984) The toxicity of sixteen metallic compounds in Chinese hamster ovary cells. Toxicol Appl Pharmacol, **74**: 330-336.

Tandon SP & Mishra MM (1969) Effect of some rare-elements on nitrification by *Nitrobacter agilis* (in liquid culture medium). Proc Natl Acad Sci India Sect A, **39**: 209-216.

Templeton DM, Paudyn A, & Baines AD (1989) Multielement analysis of biological samples by inductively coupled plasma-mass spectrometry. Biol Trace Elem Res, **22**: 17-33.

Terytze K (1987) [Entry and spread of chlorinated hydrocarbons and toxic heavy metals with reference to their quantification in the limnetic ecosystem.] Wiss Z Humboldt-Univ (Berlin) Math-Naturwiss Reihe, **36**: 860-865 (in German).

Thompson C, Dent J, & Saxby P (1988) Effects of thallium poisoning on intellectual function. Br J Psychiatry, **153**: 396-399.

Thyresson N (1950) Experimental investigation on thallium poisoning. Influence of thallium on tissue metabolism. Acta Dermato-Venereol, **30**: 417-441.

Thyresson N (1951) Experimental investigation on thallium poisoning. Acta Dermato-Venereol, **31**: 3-27.

Tikhova TS (1964) [Industrial hygiene problems in the production metallic thallium an its salts.] Gig i Sanit, **29**: 23-28 (in Russian).

Tikhova TS (1967) [Thallium and its compounds.] In: Izrael'son ZI ed. [New data on th toxicology of rare metals and their compounds.] oedinenii. Moscow, Meditsina Publishin House, pp 24-34 (in Russian).

Tjälve H, Nilsson M, & Larsson B (1982) Thallium-201: autoradiography in pigmente mice and melanin-binding *in vitro*. Acta Pharmacol Toxicol, **51**: 147-153.

Toots H & Parker RB (1977) Thallium in salt substitutes: a possible health hazard. En Res, **14**: 327-328.

Török P & Schmahl W (1982) [The environmental toxic substance thallium sulfate.] Dtsc Ärztebl, **79**(39): 31-32 (in German).

Triebig G & Büttner J (1983) [Occupational relevant neurotoxins: I. Metals and thei compounds - A literature review of the years 1970 to 1982.] Zbl Bakt Hyg I. Abt Orig B **177**: 11-36 (in German).

Trotman-Dickenson AF ed. (1973) Comprehensive inorganic chemistry: Thallium. Oxford New York, Pergamon Press, pp 1119-1172.

Truhaut R (1960) The toxicology of thallium. A summary prepared by W. B. Deichmann of Truhaut's monograph "Recherches sur la Toxicologie du Thallium." Institut National de Sécurité pour la Prévention des Accidents du Travail et des Maladies Professionelles, Paris 1959. J Occup Med, **2**: 334-336.

Tully JG, Taylor-Robinson D, Rose DL, Furr PM, & Hawkins DA (1983) Evaluation of culture media for the recovery of *Mycoplasma hominis* from the human urogenital tract Sex Transm Dis, **10**(suppl): 256-260.

Tuovinen OH & Kelly DP (1974) Studies on the growth of *Thiobacillus ferrooxidans*. IV. Influence of monovalent metal cations on ferrous iron oxidation and uranium toxicity in growing cultures. Arch Microbiol, **98**: 167-174.

UB (Federal Office for the Environment) (1992) [Pollution of allotments and kitchen gardens in former Federal territory.] In: [Data on the environment 1992/93.] Berlin, Erich Schmidt Verlag (in German).

Ueberschär K-H, Matthes S, & Vogt H (1986) [Influence of addition of thallium to food of broiler and laying hens.] In: Anke M, Baumann W, Bräunlich H, Brückner C, & Groppel B ed. [Proceedings of the 5th Symposium of Trace Elements, Jena, 14-17 July 1986.] Jena, University of Jena, pp 1233-1240 (in German).

Ulrich F & Long CNH (1955) The effects of propylthiouracil and thyrotropic hormone on the uptake of radioactive thallium by the rat thyroid. Yale J Biol Med, **27**: 371-378.

Urbain D, Reding P, Georges B, Thys O, & Ham HR (1986) The clinical value of ^{201}Tl per rectum scintigraphy in the work-up of patients with alcoholic liver disease. Eur J Nucl Med, **12**: 267-270.

US BM (US Bureau of Mines) (1985) Minerals yearbook 1985. Volume 1: Metals and minerals. Washington, DC, Government Printing Office.

US BM (US Bureau of Mines) (1989) Mineral commodity summaries 1989. Washington, DC, US Bureau of Mines/US Geological Survey.

US EPA (1978) Ambient water quality criteria: thallium. Washington, DC, US Environmental Protection Agency (PB-292444).

US EPA (1980) Ambient water quality criteria for thallium. Washington, DC, US Environmental Protection Agency (EPA-440/5-80-074).

US EPA (1986) Subchronic (90-day) toxicity of thallium (I) sulfate (Cas No. 7446-18-6) in Sprague-Dawley rats. Interim Report. Washington, DC, US Environmental Protection Agency, Office of Solid Wastes.

Valerio F, Brescianini C, Mazzucotelli A, & Frache R (1988) Seasonal variation of thallium, lead, and chromium concentrations in airborne particulate matter collected in an urban area. Sci Total Environ, **71**: 501-509.

Valerio F, Brescianini C, & Lastraioli S (1989) Airborne metals in urban areas. Int J Environ Anal Chem, **35**: 101-110.

Vandenbalck JL & Patriarche GJ (1987) Electrochemical micro-determinations of thallium(I) and chromium(VI) ions using DPASV and DP polargraphy. Sci Total Environ, **60**: 97-104.

Venugopal B & Luckey TD (1978) Metal toxicity in mammals. Volume 2: Chemical toxicity of metals and metalloids. New York, Plenum Press.

Vesterberg O, Alessio L, Brune D, Gerhardsson L, Herber R, Kazantzis G, Nordberg G, & Sabbioni E (1993) International project for producing reference values for concentrations of trace elements in human blood and urine (TRACY). Scand J Work Environ Health, **19**(suppl 1): 19-26.

Von Mühlendahl KE, Etzold R, & Krienke EG (1978) [Thallium poisonings of children.] Dtsch Med Wochenschr, **103**: 1116-1117 (in German).

Wachs B (1988) [Relevance for bodies of water of the most toxic heavy metals.] In: Bavarian Land Institute for Water Research ed. [Hazardous substances in waste water and surface water.] Munich, Oldenbourg Verlag, pp 176-243 (Report No. 42) (in German).

Wainwright AP, Kox WJ, House IM, Henry JA, Heaton R, & Seed WA (1988) Clinical features and therapy of acute thallium poisoning. Q J Med, **258**: 939-944.

Wallwork-Barber MK, Lyall K, & Ferenbaugh RW (1985) Thallium movement in a simple aquatic ecosystem. J Environ Sci Health, **20**: 689-700.

Ward JC (1930) Thallium poisoning in sheep. J Am Pharm Assoc, **19**: 556-559.

Ward JC (1931) Thallium poisoning in migratory birds. J Am Pharm Assoc, **20**: 1272-1276.

Ward JC (1936) Thallium. XIII. Symptoms and systemic action on cattle. J Am Pharm Assoc, **25**: 687-690.

Waters CB, Hawkins EC, & Knapp DW (1992) Acute thallium toxicosis in a dog. J Am Vet Med Assoc, **201**(6): 883-885.

Wei Q (1987) [Studies on spermotoxicity of thallium carbonate in drinking water and it effect on reproductive function of mice.] Chung-hua Yu Fang Hsueh Tsa Chih, 21: 141-143 (in Chinese).

Weinig E & Schmidt G (1966) [Distribution of thallium in the organism after letha thallium poisonings.] Arch Toxicol, **21**: 199-215 (in German).

Weinig E & Walz W (1971) [Distribution of thallium in kidney and liver after letha thallium poisoning.] Arch Toxicol, **27**: 217-225 (in German).

Weinig E & Zink P (1967) [Quantitative mass-spectrometrical determination of the norma thallium content in the human organism.] Arch Toxicol, **22**: 255-274 (in German).

Weisweiler W, Blome K, & Kaeding L (1985) [Quantitative determination. Thallium and lead cycles in rotary kilns for cement production.] Staub-Reinhalt Luft, **45**: 461-466 (in German).

Welz B (1983) [Atomic absorption spectrometry.] Weinheim, Verlag Chemie (in German).

Welz B, Schlemmer G, & Mudakavi JK (1988a) Investigation and elimination of chloride interference on thallium in graphite furnace atomic absorption spectrometry. Anal Chem **60**: 2567-2572.

Welz B, Schlemmer G, & Mudakavi JK (1988b) Palladium nitrate - magnesium nitrate modifier for graphite furnace atomic absorption spectrometry; part 1. Determination of arsenic, antimony, selenium and thallium in airborne particulate matter. J Anal At Spectrom, **3**: 93-97.

WHO (1986) UNEP/WHO Global Environmental Monitoring System/HEAL Project - Guidelines for integrated air, water, food and biological exposure monitoring. Geneva, World Health Organization (Document PEP/86.6).

Wiegand H (1988) Neurotoxicology of heavy metals: synaptic transmission as influenced by mono- and divalent metal cations. In: Proceedings of the 1st European Meeting of Environmental Hygiene, pp 98-102.

Wiegand H, Csicsaky M, Krämer U, Papadopoulos R, & Willeke E (1983) Neuromuscular transmission in the rat phrenic nerve hemidiaphragm preparation influenced by thallium acetate. Proceedings of the 4th International Conference on Heavy Metals in the Environment. vol 2, pp 896-899.

Wiegand H, Csicsaky M, & Krämer U (1984a) The action of thallium acetate on neuromuscular transmission in the rat phrenic nerve-diaphragm preparation. Arch Toxicol, **55**: 55-58.

Wiegand H, Papadopoulos R, Csicsaky M, & Krämer U (1984b) The action of thallium acetate on spontaneous transmitter release in the rat neuromuscular junction. Arch Toxicol, **55**: 253-257.

Wiegand H, Lohmann H, Chandra SV, & Gotzsch U (1986) The action of thallium acetate on phasic transmitter release in the mouse neuromuscular junction. Arch Toxicol, **58**: 265-270.

Wiegand H, Uhlig S, Gotzsch U, & Lohmann H (1990) The action of cobalt, cadmium and allium on presynaptic currents in mouse motor nerve endings. Neurotoxicol Teratol, **12**: 13-318.

Wiemer J, Ziskoven R, & Achenbach C (1982) Tl^+-Ions: comparison of the effects on the low inward current and contractility of ventricular tissues. Z Naturforsch, **37c**: 015-1022.

Wiemeyer SN, Jurek RM, & Moore JF (1986) Environmental contaminants in surrogates, foods, and feathers of California condors (*Gymnogyps californianus*). Environ Monit Assess, **6**: 91-111.

Williams MW, Hoeschele JD, Turner JE, Jacobson KB, Christie NT, Paton CL, Smith LH, Witschi HR, & Lee EH (1982) Chemical softness and acute metal toxicity in mice and *Drosophila*. Toxicol Appl Pharmacol, **63**: 461-469.

Wilson IC & Dean ACR (1977) Thallium sensitivity and tolerance in *Klebsiella pneumoniae*. Microbios Lett, **5**: 85-91.

Wimmer M (1985) [Mutagenic, cancerogenic, and teratogenic effects of chemical substances on man, with special reference to thallium.] Wiss Z Humboldt-Univ (Berlin) Math-Naturwiss Reihe, **34**: 789-791 (in German).

Windebank AJ (1986) Specific inhibition of myelination by lead *in vitro*; comparison with arsenic, thallium, and mercury. Exp Neurol, **94**: 203-212.

Witschi HP (1965) Desorption of some toxic heavy metals from human erythrocytes *in vitro*. Acta Haematol, **34**: 101-115.

Wood JM (1984) Alkylations of metals and the activity of metal-alkyls. Toxicol Environ Chem, **7**: 229-240.

Wood JM (1987) Biological processes involved in the cycling of elements between soil or sediments and the aqueous environment. Hydrobiologia, **149**: 31-42.

Wood JM, Cheh A, Dizikes LJ, Ridley WP, Rakow S, & Lakowicz JR (1978) Mechanisms for the biomethylation of metals and metalloids. Fed Proc, **37**: 16-21.

Woods JS & Fowler BA (1986) Alteration of hepatocellular structure and function by thallium chloride: ultrastructural, morphometric, and biochemical studies. Toxicol Appl Pharmacol, **83**: 218-229.

Wystrcil H-G, Martini M, Maier-Reiter W, & Arndt U (1987) [Ecotoxicology of thallium. Effect on photosynthesis.] Agrar-Umweltforsch Baden-Württemberg, **16**: 37-66 (in German).

Yang Q & Smeyers-Verbeke J (1991) Effectiveness of palladium matrix modification for the determination of thallium by graphite furnace atomic absorption spectrometry. Clin Chim Acta, **204**: 23-26.

Yokoyama K, Araki S, & Abe H (1990) Distribution of nerve conduction velocities in acute thallium poisoning. Muscle Nerve, **13**: 117-120.

Yopp JH, Schmid WE, & Holst RW (1974) Determination of maximum permissible level of selected chemicals that exert toxic effects on plants of economic importance in Illinois Carbondale, Illinois, Southern Illinois University (Report No. 74-33) (NTIS PB-237 654)

Zasukhina GD, Vasilyeva IM, & Sdirkova NI (1980) [Approach to the determination of a mutagenic potential of environmental pollutants as illustrated by the detection of the mutagenic effect of thallium carbonate.] Dokl Acad Nauk SSSR, **250**: 766-768 (in Russian)

Zasukhina GD, Vasilyeva IM, Sdirkova NI, Krasovsky GN, Vasyukovich LY, Kenesariev UI, & Butenko PG (1983) Mutagenic effect of thallium and mercury salts on rodent cells with different repair activities. Mutat Res, **124**: 163-173.

Zeiske W & van Driessche W (1986) Impairment of Na^+ transport across frog skin by Tl^+ effects on turnover, area density and saturation kinetics of apical Na^+ channels. Pflügers Arch, **407**: 145-152.

Zhang D (1988) [Bioassay of mutagenicity of thallium carbonate.] Huanjing Kexue, **9**: 29-32 (in Chinese).

Zhou D-X & Liu D-N (1985) Chronic thallium poisoning in a rural area of Guizhou province, China. J Environ Health, **48**: 14-18.

Ziegler E & Ziegler B (1984) [Preliminary results on the heavy metal content in hair and serum.] Fortschr Atomspektrom Spurenanal, **1**: 337-346 (in German).

Ziskoven R, Achenbach C, Bahr U, & Schulten H-R (1980) Quantitative investigations on the diaplacental transfer of thallium by field desorption mass spectrometry. Z Naturforsch, **35c**: 902-906.

Ziskoven R, Achenbach C, & Wiemer J (1982) Tl^+-Ions: influence on cardic contractility. Z Naturforsch, **37c**: 995-1005.

Ziskoven R, Achenbach C, Schulten H-R, & Roll R (1983) Thallium determinations in fetal tissues and maternal brain and kidney. Toxicol Lett, **19**: 225-231.

Zitko V (1975a) Chemistry, applications, toxicity, and pollution potential of thallium. Department of the Environment, Fisheries and Marine Service, Research and Development Directorate (Technical Report No. 518).

Zitko V (1975b) Toxicity and pollution potential of thallium. Sci Total Environ, **4**: 185-192.

Zitko V & Carson WV (1975) Accumulation of thallium in clams and mussels. Bull Environ Contam Toxicol, **14**: 530-533.

Zitko V, Carson WV, & Carson WG (1975) Thallium: occurrence in the environment and toxicity to fish. Bull Environ Contam Toxicol, **13**: 23-30.

Zook BC & Gilmore CE (1967) Thallium poisoning in dogs. J Am Vet Med Assoc, **151**: 206-217.

Zook BC, Holzworth J, & Thornton GW (1968) Thallium poisoning in cats. J Am Vet Med Assoc, **153**: 285-299.

Zyka V (1972) Thallium in plants from Alsar. Sb Geol Ved Technol Geochim, **10**: 91-96.

RESUME

1. Identité, propriétés physiques et chimiques, et méthodes d'analyse

Le thallium élémentaire est un métal mou et malléable de couleur blanc bleuâtre. Exposé à l'air humide ou à l'eau, le thallium est, selon le cas, rapidement oxydé en surface ou transformé en hydroxyde. Le thallium possède deux importants degrés d'oxydation, le thallium(I) et le thallium(III). Les composés du thallium monovalent (thalleux) se comportent comme des dérivés de métaux alcalins, par exemple du potassium, alors que les composés du thallium trivalent (dérivés thalliques) sont moins basiques et rappellent les sels d'aluminium. Contrairement aux dérivés minéraux dans lesquels l'ion thallium(I) est plus stable en solution aqueuse que l'ion thallium(III), ce dernier est plus stable dans les dérivés organiques.

Le dosage du thallium dans des échantillons provenant de l'environnement est assez difficile étant donné que la concentration est de l'ordre du µg/kg tout au plus. En général, la limite de dosage dans les minéraux, les sols et les poussières est d'environ 20 µg/kg - approximativement 0,1 µg/litre pour les solutions aqueuses - et dans les produits biologiques, de quelques µg/kg, lorsqu'on ne procède pas à une concentration préalable du thallium.

La spectrométrie d'absorption atomique avec four à tube de graphite (GF-AAS) est une méthode d'analyse qui convient bien aux applications exigeant une sensibilité élevée avec des prises d'essai réduites où la concentration en thallium est de l'ordre de quelques µg/kg. Pour obtenir une bonne précision et une bonne exactitude à des teneurs de l'ordre du µg/kg, on peut utiliser la spectrométrie de masse avec dilution isotopique (IDMS) et la spectrométrie de masse en plasma à couplage inductif (ICP-MS), avec également possibilité de dilution isotopique.

2. Sources d'exposition humaine et environnementale

La présence de thallium dans l'environnement résulte à la fois de processus naturels et de l'activité humaine. Ce métal est omniprésent dans la nature et on le rencontre en particulier dans les minerais sulfurés de divers métaux lourds, quoiqu'en principe à faibles concentrations. Il n'y a guère d'endroits où la concentration en thallium d'origine naturelle soit très élevée.

Le thallium n'est produit industriellement qu'en petites quantités (la consommation industrielle mondiale a été de 10 à 15 tonnes en 1991). Le thallium et ses dérivés ont des applications industrielles très diverses. Ils entrent dans la composition de crèmes épilatoires, de rodenticides et d'insecticides, mais ces utilisations sont désormais strictement limitées. On utilise principalement le thallium et ses dérivés dans les industries électriques et électroniques ainsi que pour la fabrication de verres spéciaux. La scintigraphie en général et le diagnostic du mélanome en particulier constituent des applications importantes du radiothallium en médecine et l'on a également recours aux dérivés d'arylthallium(III) en biochimie.

Du thallium peut être libéré dans l'environnement par des fonderies (dépôts de déchets ou émissions dans l'atmosphère), des centrales thermiques à charbon, des briqueteries et des cimenteries (dans ce cas, exclusivement des émissions dans l'atmosphère). On estime que l'industrie mobilise ainsi chaque année dans le monde de 2000 à 5000 tonnes de thallium. Les émissions de thallium dues aux différentes opérations industrielles varient largement d'un type d'industrie à l'autre.

Les émissions des centrales thermiques à charbon peuvent avoir une teneur en thallium de 700 $\mu g/m^3$ d'air et celles des cimenteries des teneurs allant jusqu'à 2500 $\mu g/m^3$. Cette dernière valeur peut être ramenée à < 25 $\mu g/m^3$ en ayant recours à d'autres matières premières et en changeant de procédé. Le thallium se volatilise lors de la combustion du charbon et des matières premières utilisées pour la production de ciment puis il se recondense à la surface des particules de cendre dans les parties de l'installation où la température est plus basse. Ces particules peuvent contenir jusqu'à 50 mg de thallium/kg de cendres volantes et sont souvent de petite taille, de sorte que seulement 50% d'entre elles sont retenues par les filtres. De même, environ un tiers des particules émises par les centrales thermiques sont de faible granulométrie et peuvent par conséquent se déposer dans les voies respiratoires inférieures.

Les effluents provenant de bassins d'évacuation de terrils et contenant dans un cas jusqu'à 1620 et dans un autre jusqu'à 36 μg/litre de thallium ont provoqué, dans les cours d'eau adjacents, l'apparition de concentrations de thallium respectivement égales à 88 et 1 μg/litre. Dans les collections d'eau de pluie situées aux alentours d'une cimenterie, on a constaté la présence de concentrations de thallium allant jusqu'à 37 μg/litre. On a trouvé dans le sol aux alentours de terrils de mines, des concentrations atteignant 60 mg/kg. Au voisinage de la base de fours à métaux, et près de briqueteries et de cimenteries, on a trouvé des concentrations respectivement égales à 2, 0,6 et 27 mg/kg de terre.

Dans les zones contaminées, la concentration de thallium dans la plupart des légumes, des fruits et des viandes est inférieure à 1 mg/kg de poids frais. Les teneurs sont plus élevées dans les choux (crucifères) avec des concentrations allant jusqu'à 45 mg/kg dans les choux frisés. Il y a corrélation entre la concentration du thallium dans les tissus des animaux d'élevage et la concentration de ce métal dans le fourrage. Au voisinage de certaines cimenteries, on a signalé des teneurs plus élevées en thallium dans le fourrage (par exemple jusqu'à 1000 mg/kg dans le colza) ainsi que dans la viande de boeuf et de lapin (jusqu'à 1,5 et 5,8 mg/kg, respectivement).

3. Transport, distribution et transformation dans l'environnement

A proximité de sources ponctuelles telles que les centrales thermiques à charbon, certaines cimenteries et fonderies de métaux, la principale source de thallium atmosphérique est constituée par les cendres volantes. Une étude a montré que près de la totalité du thallium présent dans les poussières en suspension d'une cimenterie s'y trouvait sous la forme de chlorure de thallium(I) soluble.

La destinée du thallium qui parvient jusqu'au sol (par exemple sous forme de dépôt de cendres volantes) dépend en grande partie de la nature de ce sol. C'est en particulier les sols à forte teneur en argile, en matières organiques et en oxydes de fer et manganèse qui retiennent le plus de thallium. L'accroissement de la teneur en thallium par incorporation dans des complexes stables ne se produit que dans les couches supérieures. La fixation du thallium par les végétaux est d'autant plus importante que le pH est plus bas. Dans certains sols fortement acides, d'importantes quantités de thallium peuvent parvenir par lessivage jusqu'aux eaux de surface et aux eaux souterraines.

On peut s'attendre à ce que la majeure partie du thallium présent à l'état dissous dans les eaux douces se trouve sous sa forme monovalente. Toutefois, dans les eaux où la teneur en oxygène est forte et dans la plupart des eaux de mer, il peut y avoir prédominance du thallium(III). Le thallium peut disparaître de l'eau par diverses réactions d'échange, de complexation ou de précipitation.

Le thallium peut également subir une bioconcentration, mais il est peu probable que sa concentration s'accroisse le long de la chaîne alimentaire aquatique ou terrestre.

4. Concentrations dans l'environnement et exposition humaine

Dans les secteurs qui ne sont pas contaminés, les concentrations de thallium dans l'air sont généralement inférieures à 1 ng/m^3; dans l'eau elles sont inférieures à 1 μg/litre et dans les sédiments aquatiques, inférieures à 1 mg/kg. Dans l'écorce terrestre, la concentration moyenne varie de 0,1 à 1,7 mg/kg, mais on peut trouver des concentrations beaucoup plus élevées, par exemple jusqu'à 1000 mg/kg dans le charbon et les rares minerais de thallium que l'on rencontre contiennent jusqu'à 60% de cet élément. Les produits alimentaires d'origine végétale ou animale en contiennent généralement moins de 1 mg/kg de poids sec et l'apport d'origine alimentaire moyen chez l'homme se révèle inférieur à 5 μg/jour. On estime que l'apport de thallium par la voie respiratoire est inférieur à 0,005 μg de thallium par jour.

On ne possède que des données limitées sur la concentration effective de thallium dans l'air des lieux de travail. Les valeurs les plus récemment observées (dans les années 80) étaient inférieures à 22 μg de thallium par m^3 (il s'agissait d'un four à thallium installé dans un atelier produisant un alliage spécial à base de thallium). Chez des ouvriers d'une cimenterie, on a trouvé des concentrations urinaires moyennes allant de 0,3 à 8 μg/litre et chez des ouvriers d'une fonderie, des valeurs de 0,3 à 10,5 μg/litre.

5. Cinétique et métabolisme chez les animaux de laboratoire et l'homme

La thallium est rapidement et bien absorbé au niveau des voies digestives et respiratoires et il pénètre également à travers la peau. Il se répartit rapidement dans l'ensemble des organes et traverse la barrière placentaire (comme le montre sa fixation rapide dans les tissus foetaux) ainsi que la barrière hémato-encéphalique. Comme il s'accumule rapidement dans les cellules, la concentration du thallium dans le sang total ne reflète pas sa concentration tissulaire. Dans les cas d'intoxication aiguë chez l'homme et l'animal d'expérience on constate, au début, de fortes concentrations de thallium au niveau du rein et de faibles concentrations dans les tissus adipeux et le cerveau, les concentrations dans les autres organes se situant entre les deux; ultérieurement la concentration de thallium dans le cerveau augmente également.

L'élimination du thallium peut s'effectuer par la voie digestive (essentiellement par des mécanismes indépendants de l'excrétion biliaire), par les reins, les cheveux, la peau, la sueur et le lait maternel. Il peut y avoir réabsorption intestinale, principalement au niveau du côlon avec pour conséquences une diminution de la clairance totale. Chez le rat les principales voies d'élimination sont la voie digestive (environ les deux tiers) et la voie rénale (environ un tiers); chez le lapin, ces deux voies sont d'une importance sensiblement égale. Le thallium peut également être excrété par la salive.

Comme dans le cas de nombreuses autres substances, l'excrétion du thallium n'est pas identique chez l'homme et chez l'animal de laboratoire, la vitesse d'excrétion étant généralement beaucoup plus faible chez l'homme (constante de vitesse = 0,023-0,069 jour^{-1}) que chez l'animal d'expérience (constante de vitesse moyenne = 0,18 jour^{-1}). Une autre différence importante que l'on observe entre l'homme et l'animal concerne l'importance relative des diverses voies d'excrétion. Chez l'homme, il semble que l'excrétion rénale soit beaucoup plus importante que chez l'animal, encore que sa contribution à la clairance totale n'ait pas encore été définitivement établie, principalement du fait de l'insuffisance des données. En outre, le niveau d'exposition, sa durée, les insuffisances au niveau des organes excréteurs, ainsi que le traitement anti-poison qui est administré, sont autant de facteurs qui peuvent influer considérablement sur les résultats.

Une étude a montré que la quantité totale de thallium excrétée quotidiennement était de 73% environ par la voie rénale et seulement de 3,7% environ par la voie digestive. On a estimé à 19,5% la proportion excrétée dans le système pileux et par la voie percutanée et à 3,7% la proportion excrétée dans la sueur.

Chez l'animal de laboratoire, la demi-vie biologique du thallium oscille généralement entre trois et huit jours. Chez l'homme elle est d'environ 10 jours, mais on a fait état de valeurs allant jusqu'à 30 jours.

On ne dispose d'aucune donnée sur la biotransformation du thallium.

6. Effets sur les mammifères de laboratoire et les systèmes d'épreuve *in vitro*

La toxicité des sels de thallium(I) ne varie pas de manière spectaculaire d'une espèce à l'autre. Généralement, l'ingestion d'une dose de 20 à 60 mg de thallium par kg de poids corporel est mortelle en l'espace d'une semaine. Les cobayes sont légèrement plus sensibles que les autres animaux de laboratoire. L'oxyde de thallium(III), insoluble dans l'eau, présente une toxicité aiguë par voie orale ou parentérale un peu plus faible que les sels de thallium(I). la comparaison des données de toxicité aiguë montre que le thallium présente une biodisponibilité élevée par toutes les voies d'exposition. La plupart des organes sont affectés mais on constate quelques variations intra- et interspécifiques pour ce qui concerne les signes d'intoxication et l'ordre dans lequel ils se manifestent.

Les symptômes d'une intoxication aiguë par le thallium se manifestent en général dans l'ordre suivant: on constate tout d'abord une anorexie, des vomissements et une dépression, puis apparaissent une diarrhée et des manifestations cutanées (inflammation au niveau des divers orifices, furoncles, chute de cheveux), après quoi apparaissent une dyspnée et des troubles nerveux. Enfin la mort survient par insuffisance respiratoire.

les symptômes d'une intoxication chronique sont analogues. La chute des cheveux est de règle.

L'examen histologique révèle des lésions cellulaires et en particulier une nécrose. Cette nécrose a été observée au niveau des reins, du foie, de l'intestin, du myocarde et du système nerveux. On a observé dans nombreuses cellules, un gonflement des mitochondries avec disparition des crêtes, une dilatation du réticulum endoplasmique agranulaire, une augmentation du nombre de vacuoles autophagiques et de granules de lipofuscine, enfin, une disparition des microvillosités. Les altérations fonctionnelles qu'entraîne le thallium peuvent s'expliquer par la destruction physique de la membrane des organites subcellulaires. Au niveau du coeur, les effets arythmogènes se limitent au noeud sinusal.

L'intoxication par le thallium provoque des anomalies sélectives au niveau de certains comportements, qui sont corrélées aux effets biochimiques (ce qui indique des lésions cellulaires) observés dans certaines régions de l'encéphale. Un certain nombre d'effets neurologiques semblent être dus à l'action directe du thallium, par exemple l'ataxie et les tremblements qui seraient dus à des lésions cérébelleuses ou à des modifications de l'activité endocrine provoquée par des lésions au niveau de l'hypothalamus. Le système nerveux autonome, et principalement les fibres adrénergiques, peuvent être activés par le thallium. Dans les nerfs périphériques, il semble que l'action du thallium s'exerce au niveau présynaptique, avec libération spontanée du neurotransmetteur, par antagonisme avec ce processus qui dépend du calcium.

On ne sait toujours pas exactement par quel mécanisme s'exerce la toxicité du thallium. On pense qu'il en existe plusieurs, liés les uns aux autres. Un aspect important de l'intoxication par le thallium consiste dans une augmentation sensible de la peroxydation des lipides et de l'activité d'un enzyme lysosomique, la β-galactosidase. Il en résulte un déficit en glutathion qui entraîne l'accumulation de lipides peroxydés dans l'encéphale et vraisemblablement la formation de granules de lipofuscine. Il semble que le mode d'action du thallium consiste principalement dans une perturbation de la fonction des mitochondries.

Une intoxication chronique par le thallium entraîne chez l'animal une réduction de l'activité sexuelle et les effets gonadotoxiques du thallium sont manifestes chez le mâle au niveau des organes reproducteurs. Chez des rats qui avaient reçu pendant 16 jours 10 mg de thallium par litre d'eau de boisson, on a constaté qu'au niveau des testicules, c'étaient les cellules de Sertoli qui étaient les plus sensibles et la desquamation de l'épithélium spermatogène a entraîné la présence de spermatozoïdes immatures dans le sperme. Ce phénomène pourrait expliquer le moindre taux de survie des embryons et la durée de vie réduite de la progéniture après intoxication à doses sublétales du père.

Après injection de thallium dans des oeufs de poule on a constaté chez les embryons, la présence d'effets tératogènes, une inhibition de la croissance et des anomalies dans le développement osseux; toutefois ces effets sont controversés chez les mammifères, même à des doses toxiques pour la mère. On a mis en évidence le passage transplacentaire du thallium mais les nombreuses souches de souris et de rats n'ont pour ainsi dire pas présenté d'effets tératogènes.

Deux épreuves de mutagénicité microbiologique effectuées sur *Salmonella typhimurium* ont donné des résultats négatifs et les résultats de la recherche *in vivo* d'échanges entre chromatides soeurs demeurent controversés. Toutefois, selon certaines études, on aurait observé des aberrations chromosomiques et une augmentation sensible des brèches dans l'ADN monocaténaire.

On manque d'études à long terme sur la cancérogénicité du thallium.

7. Effets sur l'homme

Etant donné que les sels de thallium sont inodores, incolores et sans saveur, leur forte toxicité et la facilité avec laquelle on pouvait s'en procurer naguère - et encore maintenant dans certains pays en développement - ont fait qu'on les a souvent utilisés à des fins de suicide, d'homicide, de tentatives d'avortement illicites avec pour résultat, dans ce cas particulier, des intoxications aiguës. D'ailleurs on considère que les intoxications par la thallium sont une des causes les plus fréquentes, à l'échelle mondiale, des empoisonnements volontaires ou accidentels chez l'homme. Ce que l'on sait des intoxications chroniques par le thallium se limite aux intoxications professionnelles, aux groupes de population vivant dans des zones contaminées et aux cas d'homicide par absorption de nombreuses petites doses.

Les symptômes de l'intoxication aiguë par le thallium dépendent de l'âge, de la voie d'administration et de la dose. Les doses qui se sont révélées mortelles varient entre 6 et 40 mg/kg, les valeurs moyennes se situant entre 10 et 15 mg/kg. Si un traitement n'est pas institué, cette dose moyenne entraîne généralement la mort en l'espace de 10 à 12 jours, mais on a également connaissance de cas où la mort est survenue en l'espace de 8 à 10 heures.

La triade gastro-entérite, polynévrite et alopécie est considérée comme le symptôme classique de l'intoxication par le thallium, mais il est arrivé qu'on n'observe ni gastroentérite ni alopécie. Il y a également d'autres symptômes qui varient dans leur séquence, leur ampleur et leur intensité.

Les symptômes de l'intoxication par le thallium sont souvent diffus et commencent par de l'anorexie, des nausées, des vomissements, une saveur métallique, de la salivation, des douleurs rétro-sternales et abdominales et quelquefois des hémorragies gastro-intestinales (selles sanglantes). On observe ensuite fréquemment une constipation qui peut être rebelle au traitement et gêner par conséquent l'action de l'antidote.

Au bout de 2 à 5 jours, les troubles caractéristiques de l'intoxication par le thallium apparaissent peu à peu, quelle soit la voie d'exposition. Les effets sur le système nerveux central et périphérique sont variables mais ce que l'on constate de manière caractéristique et systématique chez les intoxiqués par le thallium, c'est une hypersensibilité des jambes, à laquelle font suite le syndrome des "pieds brûlants" et une paresthésie. L'atteinte du système nerveux central (SNC) se traduit par des symptômes tels qu'hallucinations, léthargie, délire, convulsions et coma. Les symptômes circulatoires couramment rencontrés sont une hypertension, une tachycardie et dans les cas graves, une défaillance cardiaque. Au bout de la deuxième semaine, les cheveux commencent à tomber et quelquefois même les poils du corps. La dystrophie unguéale se manifeste par l'apparition de lunules blanches (lignes de Mee) 3 à 4 semaines après l'intoxication. Les zones noires que l'on trouve dans la papille pileuse ne sont pas dues à des dépôts de pigment ou de thallium mais à la présence de petites quantités d'air qui pénètrent dans la tige du poil.

Dans les cas mortels, la mort peut survenir quelques heures à plusieurs semaines après l'intoxication, mais la plupart du temps le décès intervient dans les 10 à 12 jours. La cause du décès est principalement due à une insuffisance rénale, neurologique et cardiaque.

En cas d'intoxication sublétale, la guérison peut souvent prendre des mois. Des séquelles peuvent subsister: problèmes neurologiques et troubles mentaux, anomalies électroencéphalographiques et cécité. En outre, il semble que les fonctions intellectuelles des survivants soient affectées.

En cas d'intoxication chronique, les symptômes sont analogues mais généralement moins prononcés qu'en cas d'intoxication aiguë. La cécité peut quelquefois être permanente. La guérison complète prend des mois et il peut y avoir des rechutes.

A l'occasion d'un incident dans une cimenterie de Lengerich en Allemagne où il y avait eu émission de thallium et que l'on a bien étudiée, on a constaté que la concentration du thallium dans le système pileux et les urines des personnes exposées n'était pas corrélée avec certaines caractéristiques généralement associées à l'intoxication chronique, mais seulement avec des symptômes neurologiques subjectifs.

L'examen de biopsies et de pièces d'autopsie après intoxication par le thallium révèle des lésions au niveau de divers organes. Par exemple, après l'ingestion de doses mortelles, on constate des hémorragies de la muqueuse intestinale, des poumons, des glandes endocrines et du coeur, des infiltrations graisseuses au niveau du foie et du myocarde, ainsi que des altérations dégénératives des glomérules et des tubules rénaux. Dans l'encéphale, on peut observer une dégénérescence graisseuse des cellules glanglionnaires, des lésions au niveau des axones et la désagrégation des gaines de myéline.

L'action directe du thallium sur le système nerveux autonome peut produire des variations dans la tension artérielle. L'intoxication par le thallium provoque une névrite symétrique périphérique. Les nerfs distaux sont davantage touchés que les nerfs proximaux et les lésions, bien que plus précoces, sont moins prononcées dans le cas des nerfs dont l'axone est court, par exemple les nerfs crâniens. Les axones sont enflés et ils présentent des vacuoles et des mitochondries distendues. Dans les cas d'intoxication mortelle, on a observé de graves lésions du nerf vague, l'énervation du sinus carotidien et des lésions au niveau des ganglions sympathiques. Dans les cas d'intoxication sublétale, les nerfs atteints peuvent subir une dégénérescence de l'axone qui peut être définitive ou s'améliorer partiellement dans les 2 ans.

Une névrite optique rétrobulbaire entraînant des troubles visuels peut s'installer et persister pendant des mois après un traitement par une crème dépilatoire à base de thallium; il peut même y avoir une atrophie du nerf optique.

On n'a guère de données concernant les effets du thallium sur la reproduction humaine. Il peut y avoir des effets indésirables sur le cycle menstruel, la libido et la puissance sexuelle masculine. On sait en outre qu'une intoxication chronique peut produire des effets sur les spermatozoïdes. Comme chez l'animal, il y a passage transplacentaire; on l'a observé dans un cas d'avortement provoqué par le thallium. Toutefois, on connaît environ 20 cas d'intoxication par le thallium en cours de grossesse où, à part le fait que les nouveaux-nés avaient un poids de naissance relativement faible et présentaient une alopécie, leur développement foetal ne paraissait pas avoir souffert.

On ne sait rien des effets cancérogènes et on ne possède aucune donnée sur les effets immunologiques du thallium. On ne possède pas non plus d'éléments suffisants en faveur d'effets génotoxiques éventuels.

Le traitement des intoxications par le thallium associe la diurèse provoquée, l'administration de charbon actif et la prévention de la réabsorption par le côlon au moyen d'administration de bleu de Prusse (hexacyanoferrate(II) de potassium et de fer(II)).

8. Relation dose-réponse chez l'homme

Dans les populations non exposées, la concentration urinaire moyenne du thallium est de 0,3 à 0,4 µg/litre. Etant donné que le thallium a une courte demi-vie biologique, exprimée en jours, si l'on suppose réunies les conditions d'un état stationnaire, la concentration urinaire peut être considérée comme un indicateur de la dose totale après absorption de thallium par inhalation ou ingestion.

On a constaté, dans un échantillon d'une population vivant à proximité d'une source d'émission de thallium dans l'atmosphère, une concentration urinaire moyenne de 5,2 µg/litre. L'existence d'une relation dose-réponse nette a été observée entre la concentration urinaire de thallium et la prévalence des effets suivants: fatigue, faiblesse, troubles du sommeil, migraines, nervosité, paresthésie, douleurs musculaires et articulaires. On a également fait état d'une relation dose-réponse analogue en utilisant la teneur des cheveux en thallium comme indicateur de l'exposition.

Le Groupe de travail a estimé qu'une exposition entraînant une concentration urinaire de thallium inférieure à 5 µg/litre n'est probablement pas susceptible de causer des effets nocifs. Dans l'intervalle 5-500 µg/litre, l'ampleur du risque et la gravité des effets nocifs restent incertaines, mais une exposition entraînant une concentration urinaire supérieure à 500 µg/litre correspond à une intoxication avec des manifestations cliniques.

9. Effets sur les autres êtres vivants au laboratoire et dans leur milieu naturel

Le thallium est nocif pour tous les organismes mais il existe des différences évidentes entre les espèces, voire entre les souches d'une même espèce. La toxicité peut être différente selon qu'il s'agit de dérivés minéraux du thallium(I) ou du thallium(III) ou encore de composés organothalliens.

L'effet le plus important du thallium sur les microorganismes semble consister dans l'inhibition de la nitrification par les bactéries terricoles. Les résultats d'une étude consacrée à ce phénomène tendent à indiquer que la structure de la communauté biologique est perturbée lorsque la concentration de thallium dans le sol se situe dans les limites de 1 à 10 mg/kg de poids sec; toutefois on n'a pas pu déterminer sous quelle forme le thallium avait été utilisé dans cette expérience.

Le thallium est fixé par toutes les parties de la plante, mais principalement par les racines. Une fois qu'il a pénétré dans la cellule, il se concentre de manière inégale dans le cytosol, probablement en étant lié à un peptide. La concentration de thallium que l'on rencontre dans les végétaux dépend des propriétés du sol (plus spécialement du pH et de la teneur en argile et en matières organiques), ainsi que du stade de développement et de la partie du végétal. Il s'accumule dans les zones contenant de la chlorophylle mais il existe des végétaux résistants au thallium où cette accumulation est moindre. La présence de thallium réduit la production d'oxygène, probablement par action directe sur les transferts d'électrons au niveau du photosystème II. La présence de chlorose traduit son effet nocif sur les pigments. En outre, il semble que la toxicité du thallium soit associée à une moindre fixation des oligo-éléments. Il y a également une action nocive sur la croissance, les racines étant plus sensibles que les feuilles ou les tiges. On a observé ces effets à dès concentrations ne dépassant pas un 1 mg de thallium par kg de tissu végétal sec, après exposition à des dérivés du thallium(I).

Dans la plupart des études consacrées aux effets du thallium sur les organismes aquatiques, on a utilisé des dérivés solubles du thallium(I). La concentration de thallium la plus faible pour laquelle on a observé des effets sur les espèces aquatiques est de 8 μg/litre, concentration qui a provoqué une réduction de la croissance de certains végétaux aquatiques. En ce qui concerne les invertébrés, les effets nocifs se font souvent sentir à des concentrations plus faibles que chez les poissons (les valeurs de la CL_{50} à 96 heures sont de 2,2 mg de thallium par litre pour les daphnies et de 120 mg/litre pour une espèce de poisson d'eau douce). La valeur la plus faible de la CL_{50} a été observée après une exposition d'environ 40 jours; elle était dans le cas de poissons, de 40 μg/litre.

Les nombreux cas d'intoxication par le thallium observés dans la faune sauvage sont dus à son utilisation à grande échelle comme rodenticide. Chez les animaux granivores et chez les prédateurs, c'est au niveau du système nerveux central ou des voies digestives que se produisent les effets les plus graves. On a également observés de tels effets chez des animaux de ferme. En outre, le thallium provoque la chute des plumes dorsales chez les canards, une salivation au niveau du nez et de la bouche chez les bovins, et une réduction de la croissance chez les poulets, les poules pondeuses, les moutons et les taureaux.

RESUMEN

1. Identidad, propiedades físicas y químicas, y métodos analíticos

El talio elemental es un metal blando y maleable de color blanco azulado. Cuando se expone al aire húmedo o al agua, se produce respectivamente una oxidación rápida de su superficie o la formación del hidróxido correspondiente. Tiene dos estados de oxidación importantes: talio(I) y talio(III). Los componentes monovalentes (talosos) se comportan de manera análoga a los metales alcalinos, como por ejemplo el potasio, mientras que los compuestos trivalentes (tálicos) son menos básicos, parecidos al aluminio. A diferencia de los compuestos inorgánicos en los que el ion talio(I) es más estable en soluciones acuosas que el ion talio(III), este último es más estable en compuestos orgánicos.

La determinación del talio en muestras del medio ambiente es algo difícil, porque sus concentraciones son del orden de µg/kg o inferiores. En general, cuando no se aplica una concentración previamente establecida de talio, los límites de la determinación en minerales, suelos y polvo son de unos 20 µg/kg, en soluciones acuosas de 0,1 µg/litro y en materiales biológicos de unos pocos µg/kg.

La espectrometría de absorción atómica en horno de grafito es un método analítico idóneo para aplicaciones en las que se necesita una alta sensibilidad debido a las pequeñas cantidades de muestra con un contenido de talio de unos pocos µg/kg. La espectrometría de masas con dilución isotópica y la espectrometría de plasma-masa con acoplamiento inductivo, posiblemente combinada con la dilución isotópica, son métodos excelentes de determinación, con buena precisión y exactitud, del orden de µg/kg.

2. Fuentes de exposición humana y ambiental

El talio está presente en el medio ambiente como consecuencia de procesos naturales y procedente de fuentes debidas a actividades humanas. Está muy extendido en la naturaleza y se encuentra sobre todo en las menas de sulfuro de diversos metales pesados, aunque suele estar en concentraciones bajas. Sólo hay unas pocas zonas con concentraciones naturales de talio muy elevadas.

La producción industrial es muy pequeña (el consumo industrial en todo el mundo en 1991 fue de 10-15 toneladas/año). El talio y sus compuestos tienen una amplia variedad de aplicaciones industriales. Ahora se ha limitado rigurosamente su uso como depilatorio humano y como rodenticida e insecticida. Sus principales aplicaciones están en las industrias eléctrica y electrónica y en la producción de vidrios especiales. Otro campo importante de aplicación es el uso de radioisótopos en medicina para la escintigrafía, así como el diagnóstico de melanomas y el uso de compuestos de ariltalio(III) en bioquímica.

Las pérdidas en el medio ambiente proceden sobre todo de la fundición de minerales (depósitos de materiales de desecho y emisiones a la atmósfera), las centrales eléctricas alimentadas por carbón, las fábricas de ladrillos y de cemento (todas ellas con emisiones a la atmósfera). Se calcula que los procesos industriales movilizan en todo el mundo de 2000 a 5000 toneladas/año. Las emisiones de talio debidas a procesos industriales varían mucho en función del tipo de industria.

Las emisiones de las centrales eléctricas alimentadas por carbón pueden contener una concentración de talio de 700 $\mu g/m^3$ de aire de salida y las de las fábricas de cemento hasta 2500 $\mu g/m^3$. Esta última cifra se puede reducir hasta < 25 $\mu g/m^3$ mediante el uso de otras materias primas y cambiando el proceso de producción. El talio se volatiliza durante la combustión del carbón o la materia prima utilizada en la fabricación de cemento y se vuelve a condensar sobre la superficie de las partículas de ceniza en las partes más frías del sistema. Estas partículas contienen hasta 50 mg de talio/kg de polvillo de ceniza y son con frecuencia de pequeño tamaño, de manera que los filtros de las fábricas de cemento retienen sólo un 50%. Alrededor de un tercio de las partículas que emiten las centrales eléctricas son también de un tamaño tan pequeño que se pueden depositar en las vías respiratorias inferiores.

Los efluentes procedentes de los depósitos de decantación de residuos mineros, con un contenido de hasta 1620 y 36 μg/litro, produjeron en los ríos de vertido niveles elevados de 88 y 1 μg/litro, respectivamente. En los estanques de agua de lluvia cercanos a una fábrica de cemento se encontraron hasta 37 μg/litro. En el suelo se han detectado concentraciones máximas de 60 mg/kg en zonas próximas a materiales de desecho de minas; en las cercanías de fundiciones de metales no preciosos y de fábricas de ladrillos y de cemento se detectaron concentraciones de 2, 0,6 y 27 mg/kg, respectivamente.

En las zonas contaminadas, la mayoría de las hortalizas, frutas y carne contienen menos de 1 mg de talio/kg de peso fresco. Las concentraciones son superiores en las coles (*Brassicaceae*), habiéndose notificado niveles de hasta 45 mg/kg en la col rizada verde. Las concentraciones de talio en los tejidos de los animales de granja se corresponden con las concentraciones en el forraje. En las cercanías de algunas fábricas de cemento, se han descrito niveles superiores en el forraje (por ejemplo, hasta 1000 mg/kg en la colza) y en la carne de vacuno y de conejo (hasta 1,5 y 5,8 mg/kg, respectivamente).

3. Transporte, distribución y transformación en el medio ambiente

Cerca de fuentes localizadas, como centrales eléctricas de carbón, algunas fábricas de cemento y operaciones de fundición de metales, la fuente principal de talio en el aire es la emisión de polvillo de ceniza. Los resultados de un estudio indican que casi todo el talio del polvo flotante procedente de una fábrica de cemento estaba presente como cloruro de talio(I) soluble.

El destino final del talio que se incorpora al suelo (debido, por ejemplo, al depósito del polvillo de ceniza) depende fundamentalmente del tipo de suelo. La retención es máxima en suelos que contienen grandes cantidades de arcilla, materia orgánica y óxidos de hierro/manganeso. La incorporación de talio a complejos estables sólo produce concentraciones más elevadas en las capas superiores del suelo. La absorción del talio por la vegetación va aumentando a medida que el pH del suelo disminuye. En algunos suelos fuertemente ácidos se puede producir lixiviación de cantidades importantes de talio al terreno y las aguas superficiales próximos.

La mayor parte del talio disuelto en agua dulce suele ser monovalente. Sin embargo, en agua dulce muy oxidada y en la mayor parte del agua marina puede predominar la forma trivalente. El talio se puede eliminar de la columna de agua y acumularse en el sedimento mediante diversas reacciones de intercambio, formación de complejos o precipitación.

Aunque puede darse una bioconcentración del talio, la bioamplificación del elemento en las redes alimentarias acuática o terrestre es improbable.

4. Niveles medioambientales y exposición humana

En zonas no contaminadas por talio, las concentraciones en el aire suelen ser < 1 ng/m^3, en el agua < 1 μg/litro y en los sedimentos del agua < 1 mg/kg. Las concentraciones medias en la corteza terrestre oscilan entre 0,1 y 1,7 mg/kg, pero es posible encontrar niveles muy elevados, por ejemplo hasta de 1000 mg/kg en el carbón, y los minerales de talio que raramente se encuentran contienen hasta un 60% del elemento. Los alimentos de origen vegetal y animal suelen contener < 1 mg/kg de peso seco y la ingestión media humana de talio con los alimentos parece ser inferior a 5 μg/día. Se estima que la absorción a través del sistema respiratorio es < 0,005 μg de talio/día.

Se dispone sólo de datos limitados sobre el contenido real de talio en el aire de los lugares de trabajo. Las concentraciones observadas más recientemente (decenio de 1980) fueron < 22 μg de talio/m^3 (en la producción de una aleación especial de talio y en una fundición de talio). El promedio de la concentración determinada en orina fue del orden de 0,3-8 μg/litro en los trabajadores del cemento y de 0,3-10,5 μg/litro en los de funderías.

5. Cinética y metabolismo en animales de laboratorio y en el ser humano

El talio se absorbe con rapidez y facilidad a través de los tractos gastrointestinal y respiratorio, así como por vía cutánea. Se distribuye en poco tiempo por todos los órganos y atraviesa la placenta (como se demuestra por la rápida absorción fetal) y la barrera hematoencefálica. Debido a su acumulación rápida en las células, las concentraciones de talio en la sangre no se corresponden con su nivel en los tejidos. En casos de intoxicación aguda de animales experimentales o de personas, se producen al principio concentraciones de talio elevadas en el riñón, bajas en el tejido adiposo y en el cerebro e intermedias en los demás órganos; luego aumentan también sus niveles en el cerebro.

La eliminación del talio se puede producir a través del tracto gastrointestinal (básicamente mediante mecanismos independientes de la excreción biliar), el riñón, el pelo, la piel, el sudor y la leche materna. Se puede producir una reabsorción intestinal (sobre todo desde el colon), con la consiguiente disminución en la eliminación total del organismo. En la rata, las principales vías de eliminación del talio son la gastrointestinal (unos dos tercios) y la renal (alrededor de un tercio), siendo semejante la contribución de ambas vías en el caso de los conejos. El talio se elimina también por la saliva.

Al igual que con otras muchas sustancias, la excreción de talio en el ser humano difiere de la observada en los animales de laboratorio; en aquél la velocidad de excreción es mucho más baja (constante de velocidad = 0,023-0,069 día^{-1}) que en éstos (la constante de velocidad media = 0,18 día^{-1}). Otra diferencia importante entre el hombre y los animales es la contribución relativa de las distintas vías de excreción. La excreción renal parece ser mucho más importante en el ser humano que en los animales, aunque no se ha determinado completamente su contribución relativa a la eliminación total del organismo, debido fundamentalmente a la falta de suficientes datos respecto al hombre. Además, los niveles de exposición, su duración, la alteración de la función de los órganos de excreción, la absorción de potasio y el tratamiento correspondiente de la intoxicación aguda pueden influir considerablemente en los resultados.

En un estudio sobre la excreción renal de talio se notificó un resultado de alrededor del 73%, mientras que a través del tracto gastrointestinal fue de sólo el 3,7% de la cantidad diaria excretada. La excreción estimada a través del pelo y la piel y del sudor fue del 19,5% y el 3,7%, respectivamente.

La semivida biológica del talio en animales de laboratorio oscila generalmente entre 3 y 8 días; en el ser humano es de unos 10 días, aunque se ha informado de valores superiores a los 30 días.

No se dispone de datos sobre su biotransformación.

6. Efectos en mamíferos de laboratorio y en sistemas de ensayo *in vitro*

No hay diferencias específicas sorprendentes por especies en cuanto a la toxicidad de las sales de talio(I). Normalmente, una ingestión oral de 20 a 60 mg de talio/kg de peso corporal es letal en un plazo de una semana. Los cobayos son ligeramente más sensibles que otros animales de experimentación. El óxido tálico(III) insoluble en agua muestra una toxicidad aguda algo más baja por vía oral o parenteral que las sales de talio(I). Al comparar los datos de toxicidad aguda se aprecia una elevada biodisponibilidad a partir de todas las vías de exposición. Afecta a la mayor parte de los órganos, pero los signos de intoxicación y la sucesión de los mismos indican una cierta variabilidad intraespecífica e interespecífica.

Los síntomas de la intoxicación aguda se suceden en general de la manera siguiente: en primer lugar anorexia, vómitos y depresión, más tarde diarrea, cambios cutáneos (inflamación en los orificios corporales, furúnculos cutáneos, pérdida de pelo) y luego disnea y trastornos nerviosos. Por último, la insuficiencia respiratoria que provoca la muerte.

Los síntomas de la intoxicación crónica son semejantes a los de la intoxicación aguda. Se produce regularmente pérdida de pelo.

En el examen histológico se puede observar necrosis u otros daños celulares. Se han detectado cambios necróticos en los riñones, el hígado, el intestino, el corazón y el sistema nervioso. En numerosas células se ha observado hinchazón de las mitocondrias y pérdida de crestas, dilataciones del retículo endoplasmático liso, aumento del número de vacuolas autofágicas y de gránulos de lipofucsina y pérdida de microvellosidades. Las alteraciones de procesos funcionales debidas al talio pueden estar provocadas por la rotura física de las membranas de los orgánulos subcelulares. En el corazón, los efectos arritmogénicos se limitan al nódulo sinoatrial.

La intoxicación por talio provoca la alteración selectiva de determinados elementos de la conducta relacionados con efectos bioquímicos (lo que indica un daño celular) en ciertas regiones cerebrales. Algunos efectos neurológicos parecen deberse a la acción directa, por ejemplo la ataxia y el temblor a causa de trastornos del cerebelo o alteraciones de la actividad endocrina debidos a cambios en el hipotálamo. El talio puede activar el sistema nervioso autónomo, fundamentalmente el adrenérgico. En los nervios periféricos parece interferir a nivel presináptico en la liberación espontánea del transmisor, ejerciendo un efecto antagónico en estos procesos dependientes del calcio.

No se conoce todavía el mecanismo exacto de la toxicidad del talio. Se han propuesto varios mecanismos, que tal vez están relacionados entre sí. Un aspecto importante de la intoxicación por talio es el aumento significativo de la peroxidación de lípidos y de la actividad de una enzima lisosómica, la β-galactosidasa. El resultado es una deficiencia de glutatión que provoca la acumulación de peróxidos de lípidos en el cerebro y, al parecer, por último, la formación de gránulos de lipofucsina. Parece que el mecanismo de acción del talio radica fundamentalmente en una alteración de la función mitocondrial.

Los animales con intoxicación crónica suelen presentar una actividad sexual reducida, y en el sistema reproductor del macho son evidentes los efectos gonadotóxicos del talio. En los testículos de ratas que recibieron 10 mg de talio/litro de agua de beber durante 16 días, las células de Sertoli fueron las más sensibles y la descamación del epitelio espermatogénico provocó la aparición de espermatozoides inmaduros en el semen. Esto podría explicar el menor índice de supervivencia de los embriones o la reducción del periodo de vida de la descendencia tras una intoxicación subletal por talio de los padres.

Tras la inyección de talio en huevos, se observaron en los embriones de pollo efectos teratogénicos, inhibición del crecimiento y trastornos del desarrollo óseo, pero en los mamíferos estos efectos son discutibles, incluso a dosis tóxicas para la madre. Aunque se ha demostrado que atraviesa la placenta, muchas estirpes de ratones y ratas no muestran efectos teratogénicos en absoluto, o sólo ligeramente.

Dos pruebas de mutagenicidad microbiológica en *Salmonella typhimurium* dieron resultados negativos, y las pruebas *in vivo* sobre intercambio de cromátides hermanas fueron controvertidas. Sin embargo, en estudios aislados se han observado aberraciones cromosómicas o un aumento significativo de la fragmentación del ADN de cadena sencilla.

No se dispone de estudios de larga duración sobre la carcinogenicidad del talio.

7. Efectos en el ser humano

Debido a que las sales de talio son insípidas, inodoras, incoloras, muy tóxicas, fáciles de obtener en el pasado e incluso ahora en algunos países en desarrollo, este elemento se ha utilizado a menudo con fines suicidas, homicidas y de aborto ilegal, provocando intoxicación aguda. Es más, se considera que la intoxicación por talio es una de las causas más frecuentes, a escala mundial, de intoxicación humana voluntaria o accidental. Los conocimientos sobre la intoxicación crónica se limitan a la exposición profesional, a grupos de población que viven en zonas contaminadas y a casos de homicidio con dosis bajas múltiples.

Los síntomas de toxicidad aguda del talio dependen de la edad, la vía de administración y la dosis. Las dosis que han resultado letales varían entre 6 y 40 mg/kg, con un promedio de 10 a 15 mg/kg. Sin tratamiento, esta dosis media suele producir la muerte en un plazo de 10 a 12 días, pero también se han descrito casos de defunción en 8-10 horas.

Se considera que la gastroenteritis, la polineuropatía y la alopecia son los tres síntomas clásicos de la intoxicación por talio, pero en algunos casos no se observó gastroenteritis ni alopecia. También se producen otros signos y síntomas, con un orden de aparición, amplitud e intensidad variables.

Los síntomas de la intoxicación son a menudo imprecisos y consisten al principio en anorexia, náuseas, vómitos, sabor metálico, salivación, dolor retrosternal y abdominal y a veces hemorragia gastrointestinal (sangre en heces). Luego se suele observar estreñimiento, que puede ser resistente al tratamiento, interfiriendo así con el antídoto administrado.

Después de un periodo de dos a cinco días aparecen lentamente algunos de los trastornos asociados normalmente al talio, con independencia de la vía de exposición. Aunque los efectos en el sistema nervioso central y periférico varían, un rasgo constante y característico de la intoxicación por talio en el hombre es la sensibilidad extrema de las piernas, a la que sigue el «síndrome de los pies urentes» y la parestesia. Su acción sobre el sistema nervioso central (SNC) se refleja en síntomas tales como alucinaciones, letargia, delirio, convulsiones y coma. Los síntomas circulatorios normales son hipertensión, taquicardia y, en los casos graves, insuficiencia cardiaca. Después de la segunda semana de la intoxicación se suele producir pérdida del pelo y, a veces, del vello; la distrofia de las uñas se detecta por la aparición de rayas semicirculares blancas (líneas de Mee) tres o cuatro semanas después de la intoxicación. Las regiones negras que se observan en las papilas pilosas no se producen por la deposición de pigmentos o de talio, sino que se deben a pequeñas cantidades de aire que entran en el tallo piloso.

En los casos letales, la muerte sobreviene en un plazo que oscila entre unas horas y varias semanas, pero normalmente se produce a los 10 ó 12 días. Las causas del fallecimiento son generalmente insuficiencia renal, del SNC y cardiaca.

En intoxicaciones subletales, la recuperación requiere con frecuencia meses. A veces persisten los trastornos neurológicos y mentales, así como las anomalías electroencefalográficas y la ceguera. Por otra parte, parece ser que los supervivientes sufren un deterioro de las funciones intelectuales.

En casos de intoxicación crónica los síntomas son semejantes, pero en general más leves que en la intoxicación aguda. A veces se produce ceguera permanente. La recuperación completa requiere meses y se puede interrumpir por recaídas.

En un caso bien investigado de emisión de talio alrededor de una fábrica de cemento de Lengerich, Alemania, las concentraciones de talio en el pelo y la orina de las personas expuestas no se correspondían con algunas características típicas que suelen estar relacionadas con la intoxicación crónica por talio, sino sólo con síntomas neurológicos subjetivos.

La autopsias y biopsias realizadas tras las intoxicaciones por talio ponen de manifiesto daños en diversos órganos. Por ejemplo, tras la ingestión de dosis letales se producen hemorragias en la mucosa intestinal, los pulmones, las glándulas endocrinas y el corazón, infiltraciones grasas en el hígado y el tejido cardiaco, así como cambios degenerativos en los glomérulos y los túbulos renales. En el cerebro se puede observar degeneración grasa de las células ganglionares, lesiones axonales y desintegración de las vainas de mielina.

Las variaciones de la presión sanguínea pueden deberse a los efectos directos del talio en el sistema nervioso autónomo. La intoxicación por talio produce neuropatía periférica mixta simétrica. Los nervios distales sufren más daños que los proximales, y los nervios de axón corto, por ejemplo los craneales, se ven afectados antes, aunque en menor grado. Los axones se hinchan y contienen vacuolas y mitocondrias dilatadas. En los casos de intoxicación letal, se han observado daños graves del nervio vago, desnervación del seno carotídeo y lesiones de los ganglios simpáticos. En la intoxicación subletal, los nervios afectados pueden sufrir degeneración axonal, con recuperación nula o sólo parcial en un plazo de dos años.

A veces se produce una neuritis retrobulbar con los consiguientes trastornos visuales, que puede persistir durante meses, después de un tratamiento con productos depilatorios con talio; este trastorno puede desembocar incluso en la atrofia óptica.

Los datos sobre los efectos del talio en la reproducción humana son limitados. Puede afectar negativamente al ciclo menstrual, la libido y la potencia masculina. Se sabe que la intoxicación crónica tiene efectos sobre el esperma. Al igual que en los estudios con animales, se ha observado que atraviesa la placenta; esto se ha puesto de manifiesto tras un aborto inducido por el talio. Sin embargo, en unos 20 casos de intoxicación por talio durante el embarazo no se vio afectado el desarrollo fetal, salvo el peso relativamente bajo y la alopecia de los recién nacidos.

No se dispone de informes sobre efectos carcinógenos o datos sobre efectos inmunológicos debidos al talio. No hay pruebas suficientes de efectos genotóxicos.

El tratamiento de la intoxicación por talio combina la diuresis forzada, el uso de carbón vegetal activado y la prevención de la reabsorción en el colon mediante la administración de azul de Prusia, hexacianoferrato(II) férrico potásico.

8. Relación dosis-respuesta en el ser humano

La concentración media de talio en orina en la población no expuesta es de 0,3 a 0,4 μg/litro. Habida cuenta de que el talio tiene una semivida biológica breve, establecida en días, y suponiendo unas condiciones estables, se puede tomar esta concentración urinaria como indicador de la dosis total tras la inhalación y la ingestión con los alimentos.

La concentración media en la orina en una muestra de población que vive cerca de una fuente de emisión de talio fue de 5,2 μg/litro. Se encontró una relación dosis-respuesta clara entre las concentraciones en la orina de talio y el predominio de cansancio, debilidad, trastornos del sueño, dolor de cabeza, nerviosismo, parestesia y dolor muscular y de las articulaciones. Se informó asimismo de una relación dosis-respuesta semejante cuando se utilizó el talio en el pelo como indicador de la exposición.

El Grupo Especial de Trabajo consideró que las exposiciones que producen concentraciones de talio en la orina inferiores a 5 μg/litro probablemente no son perjudiciales para la salud. En el margen de 5 a 500 μg/litro, la magnitud del riesgo y la gravedad de los efectos adversos son inciertas, mientras que la exposición que da lugar a más de 500 μg/litro está asociada a una intoxicación clínica.

9. Efectos en otros organismos en el laboratorio y en el medio ambiente

El talio afecta a todos los organismos, pero hay diferencias específicas de especies e incluso de variedades. Los diferentes compuestos inorgánicos de talio(I) y talio(III), así como sus compuestos orgánicos, pueden tener distinta toxicidad.

El efecto más importante del talio en los microorganismos parece ser la inhibición de la nitrificación por las bacterias del suelo. Los resultados de un estudio parecen indicar que la estructura de la flora microbiana se altera a concentraciones en el suelo comprendidas entre 1 y 10 mg/kg de peso seco, pero no se precisó la forma de talio utilizada en este experimento.

Absorben talio todas las partes de las plantas, pero sobre todo las raíces. Una vez que ha penetrado en las células, se concentra de forma desigual en el citosol, probablemente unido a un péptido. Las concentraciones de talio que se observan en las plantas dependen de las propiedades del suelo (en particular el pH y el contenido de arcilla y materia orgánica), así como de la fase de desarrollo y de la parte de la planta. Se acumula en las zonas que contienen clorofila, pero lo hace en menor grado en las plantas resistentes al talio. Reduce la producción de oxígeno, posiblemente por acción directa sobre la transferencia de electrones en el fotosistema II. Su interferencia con los pigmentos se pone de manifiesto por la aparición de clorosis. Por otra parte, en el mecanismo de la toxicidad parece intervenir una alteración de la absorción de oligoelementos. Afecta también al crecimiento, siendo más sensibles las raíces que las hojas o los tallos. Estos efectos se han descrito tras la exposición a formas monovalentes de talio con niveles de sólo 1 mg/kg de tejido vegetal seco.

En la mayoría de los estudios de los efectos en los organismos acuáticos se han utilizado compuestos solubles de talio monovalente. La concentración más baja notificada capaz de afectar a las especies acuáticas es de 8 μg/litro, con una reducción del crecimiento de las plantas. Los invertebrados se suelen ver afectados a concentraciones más bajas que los peces (los valores de la CL_{50} en 96 horas son de 2,2 mg de talio/litro para los dáfnidos y de 120 mg/litro para un pez de agua dulce). El valor más bajo de la CL_{50}, notificado tras la exposición durante unos 40 días, fue de 40 μg/litro para los peces.

Muchos casos de intoxicación por talio de la flora y fauna silvestres se han debido a su aplicación en gran escala como rodenticida. En animales que se alimentan de semillas y en depredadores afecta gravemente sobre todo al SNC y al aparato gastrointestinal. Estos mismos efectos se pueden observar en los animales de granja. A esto hay que añadir que el talio provoca una pérdida de plumas dorsales en los patos, salivación de la nariz y la boca del ganado vacuno y reducción del crecimiento de los pollos de asar, las gallinas ponedoras, las ovejas y los novillos.

Food additives and contaminants in food, principles for the safety assessment of (No. 70, 1987)
Formaldehyde (No. 89, 1989)
Genetic effects in human populations, guidelines for the study of (No. 46, 1985)
Glyphosate (No. 159, 1994)
Guidance values for human exposure limits (No. 170, 1994)
Heptachlor (No. 38, 1984)
Hexachlorobutadiene (No. 156, 1994)
Alpha- and beta-hexachlorocyclohexanes (No. 123, 1992)
Hexachlorocyclopentadiene (No. 120, 1991)
n-Hexane (No. 122, 1991)
Hydrazine (No. 68, 1987)
Hydrogen sulfide (No. 19, 1981)
Hydroquinone (No. 157, 1994)
Infancy and early childhood, principles for evaluating health risks from chemicals during (No. 59, 1986)
Inorganic lead (No. 165, 1995)
Isobenzan (No. 129, 1991)
Isophorone (No. 174, 1995)
Kelevan (No. 66, 1986)
Lasers and optical radiation (No. 23, 1982)
Lead (No. 3, 1977)[a]
Lead - environmental aspects (No. 85, 1989)
Lindane (No. 124, 1991)
Linear alkylbenzene sulfonates and related compounds (No. 169, 1996)
Magnetic fields (No. 69, 1987)
Man-made mineral fibres (No. 77, 1988)
Manganese (No. 17, 1981)
Mercury (No. 1, 1976)[a]
Mercury - environmental aspects (No. 86, 1989)
Mercury, inorganic (No. 118, 1991)
Methomyl (No. 178, 1996)
2-Methoxyethanol, 2-ethoxyethanol, and their acetates (No. 115, 1990)
Methyl bromide (No. 166, 1995)
Methylene chloride
(No. 32, 1984, 1st edition)
(No. 164, 1996, 2nd edition)
Methyl ethyl ketone (No. 143, 1992)
Methyl isobutyl ketone (No. 117, 1990)
Methylmercury (No. 101, 1990)
Methyl parathion (No. 145, 1992)
Mirex (No. 44, 1984)
Morpholine (No. 179, 1996)
Mutagenic and carcinogenic chemicals, guide to short-term tests for detecting (No. 51, 1985)
Mycotoxins (No. 11, 1979)
Mycotoxins, selected: ochratoxins, trichothecenes, ergot (No. 105, 1990)
Nephrotoxicity associated with exposure to chemicals, principles and methods for the assessment of (No. 119, 1991)
Neurotoxicity associated with exposure to chemicals, principles and methods for the assessment of (No. 60, 1986)
Nickel (No. 108, 1991)
Nitrates, nitrites, and *N*-nitroso compounds (No. 5, 1978)[a]
Nitrogen, oxides of (No. 4, 1977)[a]
2-Nitropropane (No. 138, 1992)
Noise (No. 12, 1980)[a]
Organophosphorus insecticides: a general introduction (No. 63, 1986)
Paraquat and diquat (No. 39, 1984)
Pentachlorophenol (No. 71, 1987)
Permethrin (No. 94, 1990)
Pesticide residues in food, principles for the toxicological assessment of (No. 104, 1990)
Petroleum products, selected (No. 20, 1982)
Phenol (No. 161, 1994)
d-Phenothrin (No. 96, 1990)
Phosphine and selected metal phosphides (No. 73, 1988)
Photochemical oxidants (No. 7, 1978)
Platinum (No. 125, 1991)
Polybrominated biphenyls (No. 152, 1994)
Polychlorinated biphenyls and terphenyls
(No. 2, 1976, 1st edition)[a]
(No. 140, 1992, 2nd edition)
Polychlorinated dibenzo-*p*-dioxins and dibenzofurans (No. 88, 1989)
Progeny, principles for evaluating health risks associated with exposure to chemicals during pregnancy (No. 30, 1984)
1-Propanol (No. 102, 1990)
2-Propanol (No. 103, 1990)
Propachlor (No. 147, 1993)
Propylene oxide (No. 56, 1985)
Pyrrolizidine alkaloids (No. 80, 1988)
Quintozene (No. 41, 1984)

---

[a] Out of print